AF655951

Mathematical Marvels

A Primer on

NUMBER SEQUENCES

Mathematical Marvels

A Primer on NUMBER SEQUENCES

Shailesh Shirali

Universities Press

UNIVERSITIES PRESS (INDIA) PRIVATE LIMITED

Registered Office
3-6-747/1/A and 3-6-754/1, Himayatnagar,
Hyderabad 500 029 Telangana, India
E-mail: info@universitiespress.com

Distributed by
Orient Blackswan Private Limited

Registered Office
3-6-752, Himayatnagar, Hyderabad 500 029 Telangana, India

Other Offices
Bengaluru, Bhopal, Chennai, Guwahati,
Hyderabad, Jaipur, Kolkata, Lucknow, Mumbai,
New Delhi, Noida, Patna, Vijayawada

First Published 2001
Reprinted 2004, 2010, 2014, 2017

ISBN: 978 81 7371 369 9

Typeset by
OSDATA, Hyderabad 500 029

Printed at
Bluebells Secure Prints
Hyderabad - 500018

Published by
Universities Press (India) Private Limited
3-6-747/1/A and 3-6-754/1, Himayatnagar, Hyderabad 500 029, Telangana, India
E-mail: info@universitiespress.com

Contents

Preface

This book is about number sequences: how they occur, how the patterns that lie hidden within them may be revealed to the eye, and how these patterns may be used in understanding them better. Sequences are among the most basic objects of higher mathematics, and it is highly desirable that we understand them well.

Part A, titled *The Method of Differences*, is about how to "look" at sequences, and how to analyze them. The terms *pattern* and *sequence* are explained, and a method is described for identifying the generating formula of a number sequence. The behaviour of polynomial sequences and exponential sequences is contrasted.

Part B, titled *A Gallery of Sequences*, is more in the nature of a family portrait of sequences: various sequences are studied, as one may study the families that reside in a village, or the different fauna and flora that occur within a forest, and their individual quirks and peculiarities are discussed. The sequences include the squares, cubes, primes, Fibonacci numbers, unit fractions, and more. The reader will encounter a great many unexpected facts here. After reading Part B, the reader may well feel that *every* sequence has something to offer, something that belongs uniquely to it; perhaps some numerical coincidence, some oddity, or something with deep, underlying connections to other areas of mathematics. There may, indeed, be a lot of truth behind this feeling!

We do not ask very much of the reader by way of prerequisites, other than familiarity with elementary algebra: addition, subtraction, multiplication of polynomials; concept of degree of a polynomial; the solution of simple equations; basic laws of exponents; formulas for a few expansions in algebra, . . . ; in short, the topics in algebra that are normally taught in classes VII–IX in most countries. Thus, the reader is expected to know facts such as the following:

1. basic expansions such as
   - $(n+1)^2 = n^2 + 2n + 1, \quad (n-1)^2 = n^2 - 2n + 1;$

- $(n+1)^3 = n^3 + 3n^2 + 3n + 1$;

2. basic factorizations such as
   - $a^2 - b^2 = (a-b)(a+b)$;
   - $a^3 - b^3 = (a-b)(a^2 + ab + b^2)$;
   - $a^4 - b^4 = (a-b)(a+b)(a^2 + b^2)$;
3. basic laws of exponents such as
   - $x^m \cdot x^n = x^{m+n}$, $(x^n)^2 = x^{2n}$;
   - $2^{n+1} - 2^n = 2^n$;
   - $3^{n+1} - 3^n = 2 \times 3^n$;
   - $4^{n+1} - 4^n = 3 \times 4^n$;
   - $5^{n+1} - 5^n = 4 \times 5^n$;

and so on. Logarithms make an occasional appearance and, in some places, we make use of basic facts about the prime factorization of numbers and the greatest common divisor (gcd) of two numbers. Armed with only a few such facts, it is surprising how far one can go.

Of necessity, the overall level of Part B is higher than that of Part A and will perhaps have greater appeal for the more mature reader. However, we feel that there is something in this part for everyone – even the beginner. (In some cases, the more difficult proofs have been deliberately relegated to the last section of the relevant chapter, and they may be skipped by the younger reader, if so desired.)

Many exercises will be found scattered through the book. It is important that they be tackled by the reader because, in some instances, the contents of later chapters require the results of exercises done earlier on. It would be a good idea to keep a record of discoveries made while doing the exercises.

Three appendices are included. The first is on a proof technique known as the method of mathematical induction, the second gives solutions to all the exercises in the book, and the third provides a list of references.

⋆ ⋆ ⋆

Acknowledgements The author gratefully acknowledges the help received from Dr A Kumaraswamy and Mr V Sundararaman with regard to the computer software (LaTeX) and the PCs used in typesetting this book, the feedback provided by Professor Phoolan Prasad, Dr Jayant Kirtane and Shri PK Srinivasan on the content and style of the book, and the moral support provided by his wife Padmapriya. He also thanks Universities Press for the support he has recieved from them.

⋆ ⋆ ⋆

Dedication I dedicate this book to my parents Shri Ashok R Shirali and Smt Lata A Shirali.

PART A

The Method of Differences

Chapter 1

Introduction

1.1 What is a pattern?

It is easy enough to recognize a pattern when you see one, yet, strangely, it is hard to express exactly what the word means. Most of us like patterns, for there is something regular, symmetric and orderly about them. We like to have them on our dresses, in *rangoli* patterns at the entrances of our houses, on wall-paper, on curtains, and so on. The richer and more intricate the pattern, the more we like it.

My dictionary defines "pattern" this way:

A complex of integrated parts functioning as a whole.

However, that may not make much sense to anyone! So, we try another approach. Rather than thumb through dictionaries, we shall try to understand what patterns are about by studying *examples* of patterns. We proceed to do just this, through a series of "case studies".

• **The Powers of 2** Consider the powers of 2, i.e., the numbers 2, $2^2 = 2 \times 2$, $2^3 = 2 \times 2 \times 2$, $2^4 = 2 \times 2 \times 2 \times 2$, and so on:

$$2,\ 4,\ 8,\ 16,\ 32,\ 64,\ 128,\ 256,\ \ldots.$$

If we list the last digits of these numbers, here is what we find:

$$2,\ 4,\ 8,\ 6,\ 2,\ 4,\ 8,\ 6,\ \ldots.$$

Note the pattern: the sequence 2, 4, 8, 6 repeats over and over again.

• **The Powers of 3** We do the same thing with the powers of 3, i.e., the numbers 3, $3^2 = 3 \times 3$, $3^3 = 3 \times 3 \times 3$, . . . :

$$3,\ 9,\ 27,\ 81,\ 243,\ 729,\ 2187,\ 6561,\ \ldots.$$

The last digits of these numbers are listed below.

$$3,\ 9,\ 7,\ 1,\ 3,\ 9,\ 7,\ 1,\ \ldots.$$

Once again, there is a visible and pleasing pattern: the sequence 3, 9, 7, 1 repeats over and over again.

• **Multiplication Table for 9**

Consider the multiplication table for 9:

$$\begin{aligned} 9 \times 1 &= 9, \\ 9 \times 2 &= 18, \\ 9 \times 3 &= 27, \\ 9 \times 4 &= 36, \\ 9 \times 5 &= 45, \\ 9 \times 6 &= 54, \\ 9 \times 7 &= 63, \\ 9 \times 8 &= 72, \\ 9 \times 9 &= 81. \end{aligned}$$

Study the pattern in the units and the tens digits of the products; the units digits go 9, 8, 7, 6, . . . , 2, 1, while the tens digits go 1, 2, 3, 4, . . . , 8, 9.

• **A Pattern of Squares** Consider the list of squares shown below.

$$\begin{aligned} 1^2 &= 1, \\ 11^2 &= 121, \\ 111^2 &= 12321, \\ 1111^2 &= 1234321, \\ 11111^2 &= 123454321, \\ 111111^2 &= 12345654321, \\ 1111111^2 &= 1234567654321, \\ 11111111^2 &= 123456787654321, \\ 111111111^2 &= 12345678987654321. \end{aligned}$$

There is a very appealing pattern in the numbers. Are there other lists of squares or cubes that share this curious property?

• **More multiples of 9** Another curious pattern is seen in the following table, which has numbers very similar to those seen above.

$$\begin{array}{rcccccl}
0 & \times & 9 & + & 1 & = & 1, \\
1 & \times & 9 & + & 2 & = & 11, \\
12 & \times & 9 & + & 3 & = & 111, \\
123 & \times & 9 & + & 4 & = & 1111, \\
1234 & \times & 9 & + & 5 & = & 11111, \\
12345 & \times & 9 & + & 6 & = & 111111, \\
123456 & \times & 9 & + & 7 & = & 1111111, \\
1234567 & \times & 9 & + & 8 & = & 11111111, \\
12345678 & \times & 9 & + & 9 & = & 111111111, \\
123456789 & \times & 9 & + & 10 & = & 1111111111.
\end{array}$$

Does the pattern continue beyond the last line?

• **Internal Angle of a Regular Polygon** A regular polygon is one with equal angles and equal sides, e.g., an equilateral triangle, a square, a regular pentagon, ... (For a triangle, one does not need to insist on both conditions, that is, equality of sides as well as equality of angles, because each condition implies the other.)

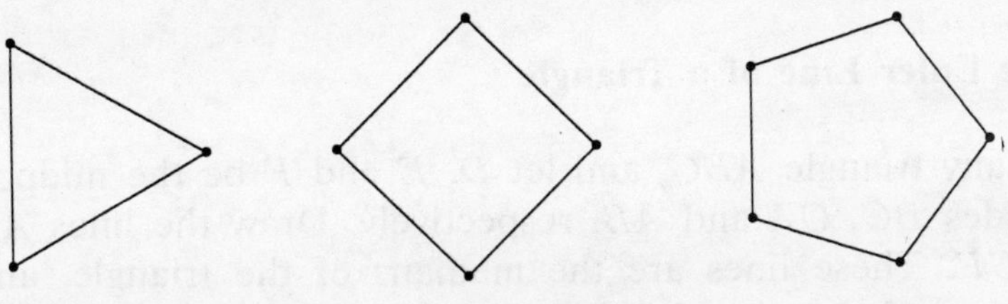

3-gon 4-gon 5-gon

Figure A.1.1 *Regular n-gons, for $n = 3$, 4 and 5*

How large is the internal angle of a regular polygon? Simple reasoning ("angle chasing") shows that the measures of the internal angles of an equilateral triangle, a square, a regular pentagon and a regular hexagon are, respectively, 60°, 90°, 108° and 120°. Is there a pattern to these numbers?

Yes! To see it, we prepare the table shown below. Here, n denotes the number of sides of the polygon ($n = 3$ for a triangle,

$n = 4$ for a square, $n = 5$ for a pentagon, ...), and A stands for the measure of its internal angle, in degrees. (Please fill in the blank slots.)

n	3	4	5	6	7	8
A	60	90	108	120		
$180 - A$	120	90	72	60		
$360/(180 - A)$	3	4	5	6		

The pattern has been exposed: $360/(180 - A)$ *is equal to* n. Once this is seen, it is rather easy to see why the pattern must hold. Try working it out for yourself. (Hint: A complete turn takes you through 360°.)

• **Kepler's Laws** Johannes Kepler was an astronomer who lived in the 1500s. He had a deep, unshakable belief that everything in the heavens had to have a pattern to it, being an expression of God, and he felt that it was his sacred duty to bring to light the hidden pattern in planetary movements. He struggled for many many years on this task, collecting enormous masses of data in the process and trying out pattern after pattern to find something that would "fit". After more than twenty years, he finally found the pattern he was looking for—and what gems the patterns turned out to be! Today, we know them as Kepler's laws of planetary motion.

• **The Euler Line of a Triangle**

Take any triangle ABC, and let D, E and F be the midpoints of the sides BC, CA and AB, respectively. Draw the lines AD, BE and CF. These lines are the medians of the triangle, and they pass through a point G—the centroid of the triangle.

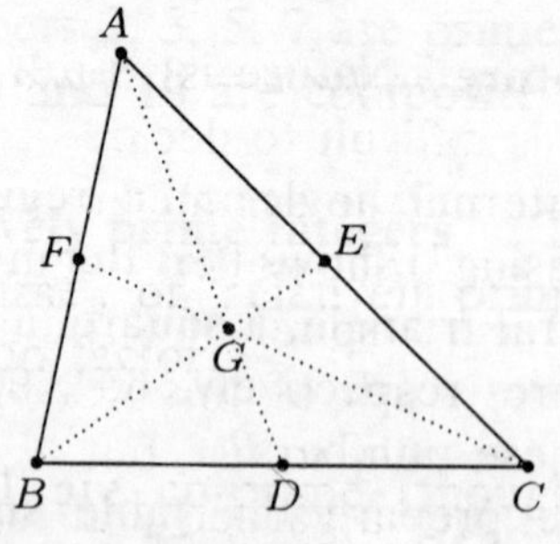

Figure A.1.2 *Centroid of a triangle*

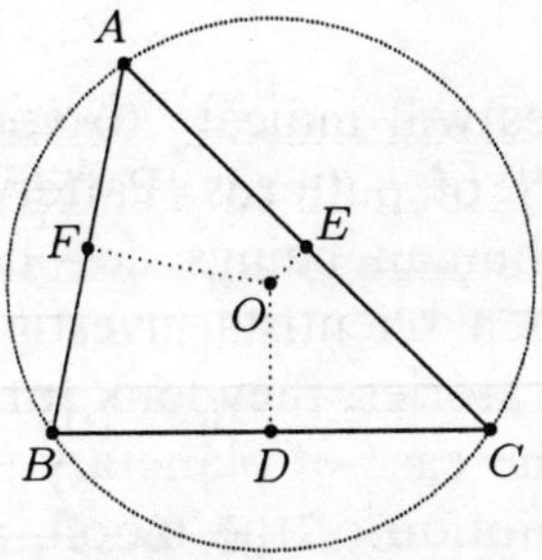

Figure A.1.3 *Circumcentre of a triangle*

Next, draw perpendiculars as follows: to BC, at D; to CA, at E; to AB, at F. These lines, which are the perpendicular bisectors of the sides of the triangle, meet in the circumcentre of the triangle – the center of the circle passing through A, B and C. We name this point O.

Finally, we drop perpendiculars from each vertex of the triangle to the opposite side: from A to BC, from B to CA, and from C to AB. These lines are the altitudes of the triangle, and they too meet in a point – the orthocenter of the triangle. We call this point H.

And now, a surprise awaits us: the points O, G and H lie in a straight line! This is the *Euler line* of the triangle, named after the great mathematician Leonhard Euler.

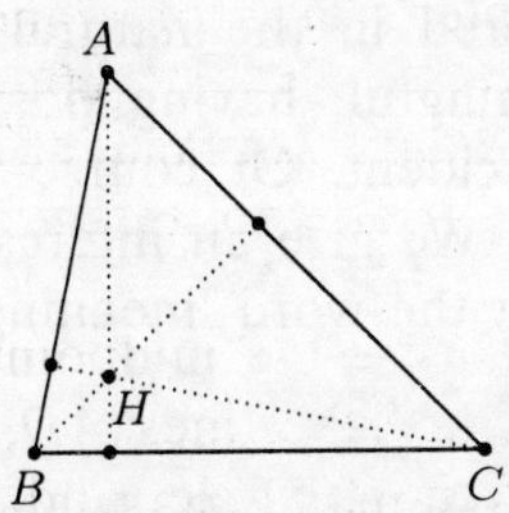

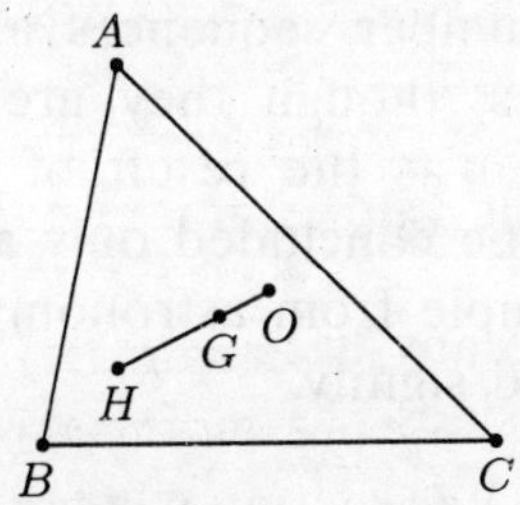

Figure A.1.4 *Orthocentre of a triangle* **Figure A.1.5** *Euler line of a triangle*

• **Examples From Nature** Nature is so full of patterns and symmetric designs that it is difficult to decide on where to start listing them! One only has to see the symmetric pattern on a butterfly's wings, or the orderly way that pollen grains are arranged within a sunflower, or the graceful spiral of a snail-shell or the horns of a blackbuck, to see the truth of this. Examples from the world of plants and flowers are easy to list, but it is simpler – and more enjoyable – to step out into a garden and look!

★ ★ ★

Perhaps these examples will indicate to you some of the beauty, variety and universality of patterns. Patterns are to be found in almost anything that human beings do—be it art, mathematics, music, or science. When scientists investigate a phenomenon in nature or in their laboratories, they look for patterns in their data, just as Kepler did in the case of planetary motion, as Newton did when he studied the motion of the moon, as Gregor Mendel did when he discovered the laws of genetics, and as Jim Watson and Francis Crick did when they discovered the structure of the DNA molecule. Any number of such examples can be listed.

In this book, we shall study patterns of a particular kind—those found in number sequences.

1.2 Number sequences

As with the word 'pattern', we shall not define the word *sequence* very precisely, as its meaning in mathematics is much the same as that in everyday life; e.g., consider the phrase, 'a sequence of events'. In mathematics too there are sequences: sequences of numbers, of points, of sets, and so on. The notion of sequence is a most vital one in advanced mathematics.

Number sequences are often encountered in the natural sciences, though they are not always meaningful, having possibly arisen as the result of coincidence or accident. Of course, this can be concluded only after careful study. We give an interesting example from astronomy to illustrate what the word 'meaningful' could signify.

• **The Titius–Bode Law** There is a law—the Titius–Bode Law—which states that the distances of the planets from the sun follow a certain pattern and can be obtained by the following arithmetical procedure. We write down the number sequence

$$0,\ 3,\ 6,\ 12,\ 24,\ 48,\ 96,\ 192,\ 384,\ \ldots,$$

where each number after the '3' is double the one to its left ($6 = 2 \times 3$, $12 = 2 \times 6$, ...). Next, we add 4 to each number and divide the result by 10. The sequence we obtain is:

$$0.4,\ 0.7,\ 1.0,\ 1.6,\ 2.8,\ 5.2,\ 10.0,\ 19.6,\ 38.8,\ \ldots.$$

According to the Titius–Bode Law, the distances of the planets from the Sun are just these numbers, the distances being expressed in 'Astronomical Units' (AU); here one AU is the distance from the Sun to the Earth. Thus, according to the law, 0.4 is the distance from the Sun to Mercury, 0.7 corresponds to Venus, 1.0 corresponds (naturally) to the Earth itself, 1.4 corresponds to Mars, and so on.

Rather surprisingly, these are exactly the observed distances! The number 5.2 corresponds to Jupiter. Historically, the number 2.8 represented a puzzle; there did not seem to be anything interesting at this distance, and there was speculation that some new planet would be found there. A major surprise came in 1781 with the discovery of Uranus at a distance of 19.2 AU, rather close to the predicted 19.6 AU! Shortly after this, in 1800, came the discovery of the asteroid Ceres – the 1st asteroid to be seen by humankind – and its distance from the Sun was found to be 2.8 AU, exactly that predicted by the Titius–Bode Law. This was a triumph for the law. Unfortunately, the next two planets (Neptune and Pluto, discovered over the next century) do not fit into the scheme at all – Neptune's predicted distance from the Sun is 38.8 AU, whereas the observed average distance is 30.1 AU, and in the case of Pluto the discrepancy is still worse: the observed average distance is 40 AU, whereas the predicted distance is 77.2 AU. So, it is possible that far from being a natural law, the Titius–Bode 'law' is really a mere coincidence (lots of them!), in which case it should not be taken too seriously. This sort of thing does happen in science every once in a while, and it illustrates our usage of the word 'meaningful'.

(Note: Some scientists have put forward the theory[1] that the law *did* once hold, with Neptune orbiting the Sun at a distance of about 38.8 AU (roughly the distance indicated by the Titius–Bode law) and Pluto a satellite of Neptune's. However, due to some reason, perhaps the influence of a star that came too close, Neptune lost its satellite. As a result their orbits changed drastically, and Pluto ended up becoming a planet.)

1.3 Sequences in mathematics

Number sequences arise in mathematics very often and in a perfectly natural manner. For instance, we have the sequence of

[1] The hypothesis is that in any planetary system, some law similar to the Titius–Bode law must hold.

squares:

$$0,\ 1,\ 4,\ 9,\ 16,\ 25,\ 36,\ 49,\ \ldots;$$

the sequence of powers of 2:

$$1,\ 2,\ 4,\ 8,\ 16,\ 32,\ 64,\ 128,\ 256,\ \ldots;$$

the sequence of primes:

$$2,\ 3,\ 5,\ 7,\ 11,\ 13,\ 17,\ 19,\ 23,\ \ldots;$$

the sequence of factorial numbers, where the nth number is the product $1 \times 2 \times 3 \times \cdots \times (n-1) \times n$:

$$1,\ 2,\ 6,\ 24,\ 120,\ 720,\ 5040,\ 40320,\ \ldots;$$

(clearly, this sequence grows with tremendous rapidity); and so on. Evidently, infinitely many such examples can be given.

Number sequences often arise as a result of some arithmetical procedure. Problems that have to do with counting often bring up extremely interesting sequences. The factorial sequence displayed above arises from such a context. Here, the nth number in the sequence is the number of different ways in which n children may be seated on a row of n chairs. Thus, three children may be seated in 6 different ways, four children in 24 different ways, and so on.

To see more clearly what is happening here, let the children be named A, B, C, ... If there are only two children (A, B), then we can arrange them in 2 ways (namely, as AB or BA). If there are three children (A, B, C) then the ways of arranging them are:

ABC, ACB, BAC, BCA, CAB, CBA,

or 6 ways in all. Likewise, with four children (A, B, C, D) we find that there are 24 different possibilities:

ABCD, ABDC, ACBD, ACDB, ADBC, ADCB, ...

List out the other ways on your own, and see for yourself that there are indeed 24 ways in all.

⋆ ⋆ ⋆

Here is another counting problem in which a sequence arises quite naturally. Let straight lines be drawn upon a flat sheet of paper with the following object in mind: *we wish to divide up the sheet into as many regions as possible*, a "region" being a portion of the paper that has been created by the lines which border the regions. If the lines are thought of as cuts made by a knife, then the regions are the pieces into which the sheet is cut. The figure shows the outcome when there are 1 or 2 lines.

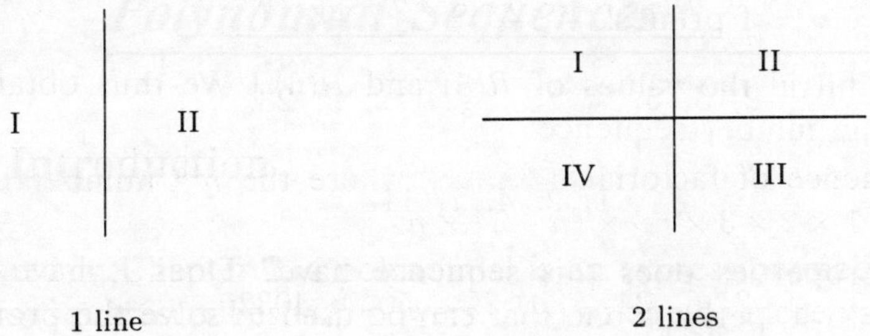

Figure A.1.6 *Regions created by drawing lines on a plane*

(By drawing the lines parallel to one another, we could create just three regions, but our intention, as stated, is to create as many regions as possible. It is easy to see that with two lines, we cannot create more than four regions.) Likewise, by drawing three lines we can create as many as seven regions.

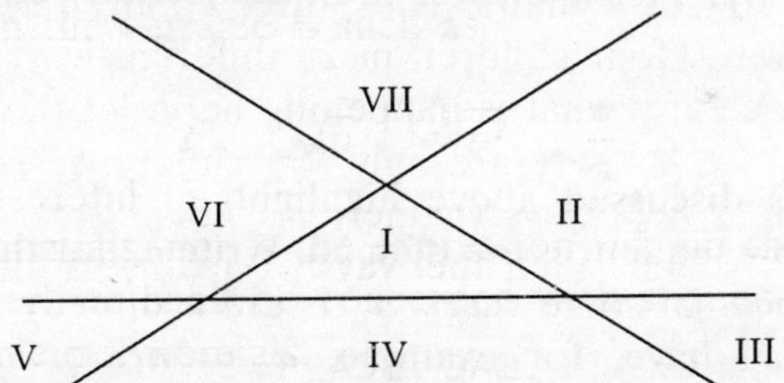

Figure A.1.7 *Regions created by three lines on a plane*

Here is the problem. Suppose that twenty lines are drawn; what is the largest number of regions that can be obtained?

We could perhaps answer the question by drawing twenty lines on a large sheet of paper and counting the regions; but this would be boring—and a waste of paper! Instead, we shall approach the problem through algebra. We introduce a symbol, $R(n)$, to denote the maximum possible number of regions when n lines are drawn ($n = 1, 2, 3, \ldots$); so $R(1) = 2$, $R(2) = 4$, $R(3) = 7$, ... We take $R(0)$ to be 1, for with no lines drawn we have just the original sheet of paper, which forms a single region. We wish to find $R(20)$. Temporarily putting aside our interest in the case $n = 20$, we proceed to list the values of $R(n)$ for smaller values of n (obtained by actual experimentation). The results are displayed below.

n	0	1	2	3	4	5	6
$R(n)$	1	2	4	7	11	?	?

[Please fill in the values of $R(5)$ and $R(6)$.] We thus obtain the following number sequence:

$$1,\ 2,\ 4,\ 7,\ 11,\ \ldots.$$

What properties does this sequence have? Does it have some pattern to it, perhaps one that can be used to solve the problem? Here is one possibility: if we compute the differences between successive numbers of the sequence, we obtain

$$2-1=1,\ 4-2=2,\ 7-4=3,\ 11-7=4,\ \ldots,$$

that is, the numbers 1, 2, 3, 4, ... This is a nice sequence to deal with! Does it continue in the manner that its pattern suggests? Assuming that it does, can we use the pattern to compute $R(20)$ or $R(100)$? Please answer the question on your own before proceeding.

⋆ ⋆ ⋆

The examples discussed above highlight an interesting question concerning how sequences are defined. Write a_n for the nth number of the sequence. In some cases, a_n is given directly as a function of n. We may have, for example, $a_n = n^2$, or $a_n = \sqrt{n}$, or $a_n = n^2 + \sqrt{n}$, or something complicated and unappealing like

$$a_n = \frac{2^n + \sqrt{n} + n - 1/n}{3^n + \sqrt[3]{n} + 1/n^2}.$$

(Probably no one would want to have anything to do with this sequence!)

In other cases, a_n is defined not directly as a function of n but *in terms of the preceding numbers of the sequence*; that is, via a "recurrence relation". We may have, for example, $a_n = a_{n-1} + 1$, or $a_n = 2a_{n-1}$, or $a_n = 2a_{n-1} + a_{n-2}$, or something complex like

$$a_n = \frac{a_{n-1}^2 + \sqrt{a_{n-2}} + 1/a_{n-3}}{a_{n-1}^3 + \sqrt[5]{a_{n-2}} + 1}.$$

In the problem concerning the regions carved out on a plane by drawing lines upon it, we had found such a relation for $R(n)$, namely,

$$R(n) = R(n-1) + n.$$

This is an example of a recursion relation. Note that it was found "experimentally".

An extremely interesting question which may be asked here is this: *If a sequence is specified by giving a formula for the* n^{th} *term, can we readily find a recursive definition? And if a sequence is specified via a recursion relation, can we readily find an "absolute" definition?* There are no clear-cut answers to these questions. All we can say is that for many of the commonly encountered sequences, both kinds of definitions are possible.

The task taken up in some detail in Chapters 3–5 deals with the second question – that of finding an "absolute" formula for the nth term, starting from the recursive definition. The strategy we shall follow is this: *given a sequence, uncover its pattern (this will typically involve recursion); use the pattern to compute more members of the sequence; and then, using the same pattern, find a formula for the* n^{th} *term.*

Described thus, the strategy is seen to be a very ancient and well-used one, used not just in mathematics but in science and astronomy too (usually with stunning effects). For instance, the dates and timings of solar and lunar eclipses can be predicted with astonishing accuracy by this technique, with an error of less than a second. Many ancient civilizations – the Mayas, the Incas – were able to do this.

The technique we use to zero in on the pattern of a sequence is essentially that of taking differences between consecutive members of the given sequence. As we shall find, the technique often yields surprising results.

Chapter 2

Terms and Symbols

In the pages to follow you will encounter many unfamiliar terms and symbols. The meanings of some of these words have been given below, for you to refer to when in doubt.

- **Sequence** A sequence is simply a list of numbers occurring in some definite order; we may also call it an "ordered set". The sequence is *finite* if it comes to a stop at some point (that is, it "terminates"); or else it is *infinite*.

Example We have the following infinite sequences:

- the sequence of natural numbers: 1, 2, 3, 4, 5, . . . ;
- the sequence of odd numbers: 1, 3, 5, 7, 9, . . . ;
- the sequence of squares: 1, 4, 9, 16, 25, . . . ;
- the powers-of-2 sequence: 1, 2, 4, 8, 16, . . . ;
- the sequence of reciprocals of the natural numbers:

$$1, \frac{1}{2}, \frac{1}{3}, \frac{1}{4}, \frac{1}{5}, \ldots.$$

The nth term of a sequence may be denoted by symbols such as a_n or b_n or c_n, with the 'n' shown as a subscript.

- **Symbol to Denote a Sequence** We often use a single symbol to denote an entire sequence. For example, we may write,

Let P represent the sequence of prime numbers 2, 3, 5, 7, 11, 13, 17, ...

Angular brackets ('$\langle$' and '$\rangle$') are often used to enclose the sequence.

Example For the sequence P referred to above, we write $P = \left\langle 2, 3, 5, 7, 11, 13, 17, \ldots \right\rangle$.

• **Recursion** When the terms of a sequence are defined using the preceding terms, then the sequence is said to be defined *recursively*.

Example Consider the powers-of-2 sequence. Let a_n denote the nth number in the sequence. Each number in the sequence is twice the preceding one, so we have the following recursion:

$$a_n = 2a_{n-1} \qquad \text{(valid for all } n > 1\text{)}.$$

• **Variable** A *variable* is a quantity that changes, as a result of which other quantities that depend on it also change. Quantities that do not change, that is, whose values are fixed, are referred to as *constants*.

Example Let m and n be quantities connected by the relation $m = 2n + 3$. Then as n takes the values 0, 1, 2, 3, 4, ..., m correspondingly takes the values 3, 5, 7, 9, 11, ... Both m and n are variables.

The idea of a constant is more easily grasped when one studies physics. For example, the charge on an electron is a constant, as is the mass of an electron or proton. The speed of light, c, is a constant for a given medium. The acceleration due to gravity, g, is a constant for each fixed location on the Earth (roughly $9.8\,\text{m/s}^2$); so is Newton's gravitational constant, G. On the other hand, consider a simple pendulum. Its length l can be changed, and this leads to a change in t, its period of oscillation; so l and t are variables. It can be shown that $t = 2\pi\sqrt{l/g}$.

In a lighter vein, for tourists coming to India, cows seated sedately on the medians of roads and buffaloes sauntering across the road are constants, and train timings are variables!

⋆ ⋆ ⋆

It is common in mathematics to use symbols such as u, v, w, x, y, z, ... when we refer to variables. (Greek letters are used too – but I shall not frighten you with these!) When we refer to variables that take only integer values (0, ±1, ±2, ...), we often use symbols such as i, m, n. On the other hand, when the variables are allowed to take fractional values as well (0.1, 1.32, 2.739, ...), we usually use symbols such as w, x, y, z, and so on.

For constants, the symbols commonly used are a, b, c, and so on.

An expression made up of one or more variables is usually given a name; for example, we may give the name f to the expression $x^2y + z^4$. As the variables involved are x, y and z, we often write the name as $f(x, y, z)$ to draw attention to the variables that occur in the expression.

Example The expression $w + 1/x + y^z$ has four variables, and $kl^3 + mn + p/q^2$ has six variables.

• **Monomial Expression** Expressions such as x, yz, ab^2, $-2ab$, $9cd$, ..., where variables and numbers are multiplied together to make a single quantity, are called *monomials*.

Example The expressions

$$x^2,\ 3x^3y,\ 8x^4y^7,\ -4mnp^3,\ -13a^3b,\ \ldots,$$

are monomials. The expression $a - 2b$ is not a monomial.

• **Binomial Expression** A sum of two monomials is a *binomial* expression, and a sum of three monomials is a *trinomial* expression.

Example The expressions $a + 2b$, $3b + 5c^2d$, $7x - yz$, $m + 2ab$, ..., are binomial expressions. An example of a trinomial expression is $x + 2y - 3z$.

• **Polynomial Expression** A sum of monomials is a *polynomial* expression; thus, a polynomial is a sum of one or more terms, each of which is a product of numbers and variables. (The word *poly* comes from Latin, and means 'many'.) Note that the exponents occurring in a polynomial are all whole numbers.

Example 1 $P(x) = x^7 - 8x^5 + 3x + 10$ is a one-variable polynomial (in the variable x).

Example 2 $Q(x, y) = x^3y^2 - 10xy^4 + 7y^5$ is a two-variable polynomial (in the variables x and y).

Example 3 The expression $R(a, b, c) = a + 2bc - ac$ is a three-variable polynomial (in the variables a, b and c.)

Example 4 The expression $A(x) = \sqrt{x}$ is *not* a polynomial, as the exponent of x in $A(x)$ is not a whole number. Likewise, $1/x$, $\sqrt{x^5}$ and $\sqrt[5]{x^2}$ are not polynomials.

• **Degree of a Polynomial in One Variable** The *degree* of a one-variable polynomial P is the largest exponent occurring in P.

Example 1 The degree of $4x^2 + 3x$ is 2.

Example 2 The degree of $-y^3 + 4y^2 + 1$ is 3.

Example 3 The degree of $5m^7 + 6m^5 + 3$ is 7.

Example 4 The degree of $z^{101} + z^{37} + 1$ is 101.

• **Polynomial Sequence** A sequence generated by a one-variable polynomial, by substituting the values 0, 1, 2, 3, 4, ..., successively in place of the variable, is called a *polynomial sequence*.

Example The polynomial $R(n) = n^4 - 7n$ generates the sequence

$$\left\langle 0,\ -6,\ 2,\ 60,\ 228,\ 590,\ 1254,\ 2352,\ \ldots \right\rangle.$$

(Note that we have used the symbol n rather than x for the variable, because of the restriction to integral values.)

• **Exponential Expression** An expression in which the variable occurs as an exponent is called an *exponential expression*.

Example 1 2^x, $2^{x(x+1)}$ and 3^{x^2} are one-variable exponential expressions in the variable x.

Example 2 $2^x 5^{-y}$ and $3^{x^2} 7^{y+1}$ are two-variable exponential expressions in the variables x and y.

Example 3 $2^x + x^3$ is a sum of both types of expressions.

• **Exponential Sequence** A sequence generated by a one-variable exponential expression, by substituting the values 0, 1, 2, 3, 4, ..., successively in place of the variable is called an *exponential sequence*.

Example The exponential expression $A(n) = 3^n - 2^n$ generates the sequence

$$\left\langle 0,\ 1,\ 5,\ 19,\ 65,\ 211,\ 665,\ 2059,\ \ldots \right\rangle.$$

• **Function** We shall not use the terminology of function extensively in this book, yet the concept permeates most of what we plan to do, so for the sake of completeness we must give a comprehensive definition. Let A and B be sets, A being called the *domain* and B the *co-domain*. A rule which assigns to each individual element in A some element in B, is called a *function from A to B*. For example, we may assign to each word in the English language its 1st letter. Here, the domain is the set of words in the English language and the co-domain the set of letters of the alphabet, namely $\{a, b, c, d, \ldots, x, y, z\}$. Or we may assign to each number its square. If we call the function f, then we write for short $f(x) = x^2$. Here, the domain and co-domain can both be taken to be the set of real numbers, $\mathbf{R}$. Or we may assign to each positive integer the sum of the divisors of that number; eg, to 10 we assign the number 18, because $1+2+5+10 = 18$. Here, the domain and co-domain can both be taken to be the set of natural numbers, $\mathbf{N}$.

• **Constant Sequence** A sequence in which the entries are all equal to one another is called a *constant sequence*.

Example The sequence $\left\langle 2, 2, 2, 2, 2, 2, 2, 2, \ldots \right\rangle$.

• **Sequence of 1st Differences** Given a sequence S, say,

$$S = \left\langle a_1,\ a_2,\ a_3,\ a_4,\ a_5,\ \ldots \right\rangle,$$

the sequence $D(S)$ of differences which is computed as

$$D(S) = \left\langle a_2 - a_1,\ a_3 - a_2,\ a_4 - a_3,\ a_5 - a_4,\ \ldots \right\rangle,$$

is the *sequence of 1st differences* of S. We use the symbol '$D(S)$' as a reminder to say that the sequence has been generated by the 1st differences of S ('D' denoting 'difference'). Sometimes we use the symbol $D_1(S)$ instead of $D(S)$. (The reason for this will soon become clear.)

Example 1 If S is given by

$$S = \left\langle 1,\ 4,\ 7,\ 15,\ 28,\ 47,\ 81,\ 119,\ 162,\ \ldots \right\rangle,$$

then the sequence of 1st differences of S is

$$D(S) = \left\langle 3,\ 3,\ 8,\ 13,\ 19,\ 34,\ 38,\ 43,\ \ldots \right\rangle.$$

Example 2 If S is

$$\left\langle 1, \ -2, \ 3, \ -4, \ 5, \ -6, \ 7, \ -8, \ \ldots \right\rangle,$$

then the sequence of 1st differences of S is

$$D(S) = \left\langle -3, \ 5, \ -7, \ 9, \ -11, \ 13, \ -15, \ \ldots \right\rangle.$$

• **Sequence of 2nd Differences** Let S be any sequence of numbers and let $D(S)$ be its sequence of 1st differences. Then, the sequence of 1st differences of $D(S)$ is the *sequence of 2nd differences* of S. We use the symbol $D_2(S)$ for this sequence. (We could also use the symbol $DD(S)$, but $D_2(S)$ seems more convenient.)

Example For the sequence S given in Example 1 above,

$$S = \left\langle 1, \ 4, \ 7, \ 15, \ 28, \ 47, \ 81, \ 119, \ 152, \ \ldots \right\rangle,$$

the sequence of 2nd differences is

$$D_2(S) = \left\langle 0, \ 5, \ 5, \ 6, \ 15, \ 4, \ 5, \ \ldots \right\rangle.$$

• **Hypothesis** Sometimes scientists and mathematicians have to make guesses based on their observations. A "guess" of this sort, when based upon carefully recorded observations and analysis, is called a *hypothesis*; sometimes we may call it a *theory*. Usually, it is put forward to explain some observed phenomenon.

Example Some of the nicest examples come from astronomy; the Titius-Bode "law" mentioned earlier, in Chapter 1, is one such. Here are a few others.

In the mid-1800s, the French astronomer-mathematician Urbain Leverrier[1] suggested the possibility of an extra planet beyond Uranus. This was done to explain some peculiarities noticed in Uranus' orbit, the idea being that the 'peculiarities' were being caused by the gravitational pull of some unseen planet. Amazingly, a new planet was found – and just where Leverrier said it would be! This is the planet Neptune, and its discovery represents one of the great success stories in mathematical astronomy. It turned out,

---

[1] The English mathematician John Couch Adams had the same idea, but no one listened to him.

however, that Neptune could not account for all the oddities of Uranus' orbit, and ultimately another such hypothesis was offered. That is how Pluto was discovered.

* * *

Here is another example. Some astronomers have suggested that the Sun has a companion star which periodically comes rather close to the Sun – every 26 million years or so! As a result of the 'close encounter', some debris is dislodged from the so-called "Oort cloud", a gigantic cloud of ice and dust and rocks that surrounds the entire Solar System at a very great distance. The debris falls towards the Sun in the form of comets, and once in a while these comets collide with the Earth. The result is a catastrophic explosion, and large-scale extinctions occur on Earth. This hypothesis has been put forward to explain the periodic large-scale extinctions that have indeed occurred in the past. Rather fittingly, the yet-to-be-observed star has been called Nemesis![2] Of course, apart from this rather indirect evidence, no one has any idea where Nemesis might be located, or even whether it exists at all.

* * *

Some theories die out naturally, because things do not quite work out as predicted by the theory. Back in the Middle Ages, the great astronomer Johannes Kepler put forward a rather fantastic and complicated theory about the Solar System. It sought to connect the Solar System with the so-called "Platonic solids", which are five in number: the cube, tetrahedron, octahedron, icosahedron and dodecahedron, and it went roughly like this. Saturn was the outermost planet known at the time, and Kepler mentally constructed a sphere that would just hold its orbit. In this sphere he inscribed a cube, and in the cube a 2nd sphere. ("Inscribed" means that the cube fits into the sphere exactly; likewise, the 2nd sphere fits exactly into the cube.) This sphere held the orbit of Jupiter. In this sphere he inscribed a tetrahedron, and in the tetrahedron a 3rd sphere. This sphere held the orbit of Mars. In this sphere he inscribed a dodecahedron, and in the dodecahedron a 4th sphere. This sphere held the orbit of the

[2] Greek goddess of Retribution. Daughter, according to Hesiod, of Night.

Earth. And so on; this continued down till Mercury – six spheres in all. This was rather convenient, because the number of Platonic solids available (namely, five) was just right! It follows that in this theory, there is no room for any further planets. What happened when the outer planets (Uranus, Neptune, . . .) were discovered? The theory died a quiet death and, in time, was forgotten . . . .

⋆ ⋆ ⋆

Two more further illustrations of "numerical hypotheses" are given below.

Example 1 Given only the 1st five terms of a sequence A, where

$$A = \left\langle 0,\ 1,\ 4,\ 9,\ 16,\ \ldots \right\rangle,$$

we notice that the numbers are all squares. We may, therefore, hypothesize or guess that the sequence is generated by the polynomial $P(n) = n^2$.

Example 2 Given only the 1st five terms of a sequence B, where

$$B = \langle 0,\ 2,\ 12,\ 36,\ 80,\ \rangle,$$

we may, on noticing that $12 = 2^3 + 2^2$, $36 = 3^3 + 3^2$, . . . , propose the hypothesis that the sequence is generated by the polynomial $Q(n) = n^3 + n^2$.

⋆ ⋆ ⋆

• **Symbols for Sequences** As noted above, we sometimes use a symbol or name to denote an entire sequence. For example, we may say,

Let P represent the sequence of primes.

Sometimes, however, we may actually possess the *generating formula* of the sequence; i.e., an expression in some variable, say n, such that when we substitute the values $n = 0, 1, 2, 3, \ldots$ in place of the variable, we generate the entire sequence. When this happens, it is convenient to choose the name so that the generating formula can be "seen" in the name. For instance, a convenient name for the sequence of cubes,

$$\langle 0,\ 1,\ 8,\ 27,\ 64,\ 125,\ 216,\ 343,\ 512,\ 729,\ \ldots \rangle,$$

is $\{n^3\}$. Likewise, a convenient symbol for the sequence of powers of 2,

$$\langle 1,\ 2,\ 4,\ 8,\ 16,\ 32,\ 64,\ 128,\ 256,\ 512\ldots \rangle,$$

is $\{2^n\}$. For the sequence

$$\langle 0,\ 2,\ 6,\ 12,\ 20,\ 30,\ 42,\ 56,\ 72,\ 90,\ 110,\ \ldots \rangle,$$

the symbol $\{n(n+1)\}$ is appropriate, as the numbers of the sequence are

$$0 \times 1,\ 1 \times 2,\ 2 \times 3,\ 3 \times 4,\ 4 \times 5,\ 5 \times 6,\ \ldots.$$

The sequence

$$\langle 0,\ 1,\ 2,\ 3,\ 4,\ 5,\ 6,\ 7,\ 8,\ 9,\ \ldots \rangle$$

would thus be denoted by $\{n\}$, and the constant sequence

$$\langle 1,\ 1,\ 1,\ 1,\ 1,\ 1,\ 1,\ \ldots \rangle,$$

(in which every entry is 1) by $\{1\}$.

• **Prime number** An integer $n \neq 1$ is said to be *prime* if it has no divisors other than 1 and n itself, and *composite* if it does have divisors other than 1 and n. 1 itself is a *unit*.

Example The numbers 2, 3, 5, 7 are prime, and so are 127, 257 and 1999; but 6, 8, 9 and 10 are composite

• **Co-prime or relatively prime integers** Two integers a and b are said to be 'co-prime', or 'relatively prime to one another', if they share no common factor.

Example 10 and 21 are co-prime (though neither number is prime), and so are 30 and 77; but not 25 and 35, which share the common factor 5; nor 22 and 24, which share the common factor 2.

• **Arithmetic progression** An arithmetic progression, or AP for short, is an infinite sequence of numbers in which the difference between consecutive terms stays constant.

Example The sequence of odd numbers

$$1,\ 3,\ 5,\ 7,\ 9,\ \ldots$$

is an AP, and so is the sequence of numbers of the form $4k+1$:

$$1,\ 5,\ 9,\ 13,\ 17,\ \ldots$$

The general AP may be written as

$$a,\ a+d,\ a+2d,\ a+3d,\ a+4d,\ \ldots.$$

Here d is called the *common difference* of the AP, while a is called the *first term*.

• **Rational number** A number r is said to be "rational" if it can be written in the form a/b, where a and b are integers, with $b > 0$.

Example The numbers 1/2, 2/3, 4/7, 5/11, ... are rational.
The decimal fraction expansion of a rational number either terminates, or else it eventually recurs. For example,

$$\begin{aligned}
1/2 &= 0.5 && \text{(terminates);}\\
7/32 &= 0.21875 && \text{(terminates);}\\
2/3 &= 0.666666666666\ldots && \text{(recurs);}\\
4/7 &= 0.571428\,571428\ldots && \text{(recurs);}\\
5/11 &= 0.45\,45\,45\,45\,45\ldots && \text{(recurs);}\\
7/30 &= 0.2\,33333333333\ldots && \text{(eventually recurs);}\\
611/4950 &= 0.12\,34\,34\,34\,34\ldots && \text{(eventually recurs).}
\end{aligned}$$

Remark "Termination" may be considered as a special case of "eventual recurrence"; for example, we have $1/2 = 0.5000\ldots$ and $7/32 = 0.21875000\ldots$.

• **Irrational number** A number is said to be "irrational" if it cannot be written in the form a/b, where a and b are integers, with $b \neq 0$.

Example Many examples of irrational numbers can be given, but in each case it is by no means obvious why they should be labelled as irrational. Thus, the numbers

$$\sqrt{2}, \quad \sqrt{3}, \quad \sqrt{5}, \quad \ldots, \quad \pi, \quad \log_2 10, \quad \ldots$$

are all irrational. This means that their decimal expansions neither terminate, nor do they recur; they are infinite and non-terminating.

Proving that the numbers listed above are irrational requires some effort and, in the case of π, the effort required is considerable!

• **Partial sums of a sequence** Let a_1, a_2, a_3, a_4, ..., be a sequence of numbers. The sequence

$$a_1, \quad a_1 + a_2, \quad a_1 + a_2 + a_3, \quad a_1 + a_2 + a_3 + a_4, \quad \ldots,$$

obtained by cumulatively adding its terms, is referred to as its "sequence of partial sums".

Example The sequence of partial sums of $\{n\}$ (the sequence of natural numbers 1, 2, 3, 4, ...) is

$$1, \ 3, \ 6, \ 10, \ 15, \ \ldots,$$

(this is the sequence of "triangular numbers"), and the sequence of partial sums of $\{2n-1\}$ (the sequence of odd numbers 1, 3, 5, 7, 9, ...) is

$$1, \ 4, \ 9, \ 16, \ 25, \ \ldots,$$

which is the sequence of squares, $\{n^2\}$.

• **Induction** This is a proof technique often used for proving statements made about the positive integers.

Let $\mathcal{P}(n)$ refer to a property of the positive integer n which, we suspect, is true for all values of n. We wish to prove this. The method of induction consists of the following steps:

(a) Show that $\mathcal{P}(1)$ is true. This is called "anchoring" the induction.

(b) Show that if $\mathcal{P}(n)$ is true, then so is $\mathcal{P}(n+1)$. This is called the "inductive step".

If these two steps are successfully accomplished, then $\mathcal{P}(n)$ has been shown to be true for all positive integers n.

The technique may be likened to climbing a ladder. Step (a) is like getting on the 1st rung, and step (b), the inductive step, is the act of stepping from one rung of the ladder to the next one.

Sometimes we take the anchor to be an integer greater than 1, for example, in the following proposition: *The inequality* $2^n > n^3$ *holds for positive integers* $n \geq 10$. Here, we would first establish $\mathcal{P}(10)$ and only then proceed to the inductive step.

- **Base-10 integer** This refers to the normal decimal representation of an integer.

Example The number 234 is actually a short form for the quantity

$$2 \times 10^2 + 3 \times 10^1 + 4 \times 10^0.$$

Historically, this kind of representation was first used by the ancient Indians. The idea reached Europe via the Arab traders, and it took root only after two centuries or so, through the pioneering efforts of mathematicians such as Leonardo of Pisa (better known as Fibonacci), L Pacioli and S Stevin.

- **Base-2 integer** This refers to the representation of an integer as a sum of distinct non-negative powers of 2.

Example The number $(10101)_2$ refers to the quantity

$$1 \times 2^4 + 0 \times 2^3 + 1 \times 2^2 + 0 \times 2^1 + 1 \times 2^0 = (25)_{10},$$

that is, the number 25 in normal decimal notation.

- **Base-9 integer** This refers to the representation of an integer as a sum of distinct non-negative powers of 9.

Example The number $(2801)_9$ refers to the quantity

$$2 \times 9^3 + 8 \times 9^2 + 0 \times 9^1 + 1 \times 9^0 = (2107)_{10},$$

that is, the number 2107 in normal decimal notation.

⋆ ⋆ ⋆

The definitions that follow are needed for some of the exercises in Chapter 5.

• **Collinearity** Points which lie on a straight line are said to be *collinear*.

Example In any triangle, the circumcentre, centroid and orthocenter are collinear. (They lie on the "Euler line", as described in the Chapter 1.)

• **Concurrence** Lines which pass through a common point are said to be *concurrent*.

Example The diameters of a circle concur (at its center), and the medians of a triangle concur (at its centroid).

Chapter 3

Polynomial Sequences

3.1 Introduction

As stated in Chapter 2, a polynomial is an expression made up of powers (squares, cubes, 4th powers, ...) of one or more variables, and cross-products of the same powers. For example, expressions such as

$$n^2 + 3n + 1, \quad n^9 - 1, \quad 2n^{10} - 6n^2 + 1$$

are polynomials in the variable n. Likewise, expressions such as

$$3x^4y - 7xy^3 + 8y^7, \quad x^8y^9 + 1, \quad x^{10}y^2 - y$$

are polynomials in the variables x and y. On the other hand, expressions in the variable n such as

$$2^n, \quad \sqrt{n}, \quad \sqrt[3]{n^2}, \quad \frac{1}{n}$$

are *not* polynomials, nor is the expression m^n (in the variable n). In this chapter, we shall be concerned with sequences generated by one-variable polynomials.

To start with, we take the sequence of squares of whole numbers, which we denote symbolically by $\{n^2\}$. As noted earlier, the notation is a natural one because the formula within the brackets is the generating formula for the sequence. Here is the sequence:

$$\left\langle 0,\ 1,\ 4,\ 9,\ 16,\ 25,\ 36,\ 49,\ \ldots \right\rangle.$$

(For reasons of convenience, we choose to start the sequence from 0 rather than from 1. The reasons for this choice will become clearer as we go on.)

The sequence of 1st differences of this sequence is defined to be the following sequence:

$$\langle 1-0=1,\ 4-1=3,\ 9-4=5,\ 16-9=7,\ \ldots \rangle,$$

that is, its terms are the differences between successive members of the original sequence. We refer to this sequence as the *sequence of 1st differences of the original sequence.* In this case, the sequence of 1st differences is

$$\langle 1, 3, 5, 7, 9, \ldots \rangle.$$

Note the very visible pattern in this sequence.

The normal procedure is to write down this sequence in a row below the original sequence; the two sequences then form an array as displayed below.

0		1		4		9		16		25		36		49
	1		3		5		7		9		11		13	

(Note the manner in which the 2nd row has been written: the 2nd row has been indented a bit, the '1' being placed between the '0' and '1' of the 1st row, and so on.) Do you notice the pattern? *The sequence of 1st differences consists of just the odd numbers, arranged in increasing order!*

If we take the differences of this sequence, i.e., of $\langle 1, 3, 5, 7, \ldots \rangle$, we obtain the sequence of 2nd differences $3-1=2$, $5-3=2$, $7-5=2$, ..., which we write in a row form below the sequence of 1st differences as shown below.

0		1		4		9		16		25		36		49
	1		3		5		7		9		11		13	
		2		2		2		2		2		2		

The pleasing discovery now is that the sequence of 2nd differences consists entirely of 2s. *We have identified a characteristic property of the square-number sequence.*

Exercises

3.1.1 Obtain the sequences of 1st and 2nd differences for the sequence $\{n(n+3)\}$:

$$\langle 0, 4, 10, 18, 28, 40, 54, \ldots \rangle.$$

What do you find?

3.1.2 Obtain the sequences of 1st, 2nd and 3rd differences for the cubic-number sequence $\{n^3\}$:

$$\langle 0, 1, 8, 27, 64, 125, 216, 343, \ldots \rangle.$$

What is the result?

3.1.3 Repeat the experiment with the sequence $\{n^4\}$, i.e., the sequence of 4th powers:

$$\langle 0, 1, 16, 81, 256, 625, 1296, \ldots \rangle.$$

Construct the successive difference sequences (1st, 2nd, 3rd, ...), till a stage is reached when the differences are constant. At which stage does this happen?

3.1.4 Repeat the experiment with the sequence $\{n^5\}$:

$$\langle 0, 1, 32, 243, 1024, \ldots \rangle.$$

(Be prepared for a lot of calculation in this exercise!) At which stage do the differences become constant?

3.1.5 On the basis of your results, what conclusion can you draw? Combine the various results and make an informed guess about the pattern.

Comment on "guess-work"

A guess of the type referred to in Problem 3.1.5 is called a *hypothesis*. It is important to realize that a hypothesis is not necessarily correct, even though it may seem to fit the given situation perfectly. This may seem odd at first sight, yet it happens in mathematics and the sciences quite often: hypotheses are put forward that seem to be correct, yet they are later disproved by experimentation. Here is a famous example of this. The amateur-mathematician Pierre de Fermat—by profession a jurist!—once claimed that the numbers

$$2^{2^0}+1, \quad 2^{2^1}+1, \quad 2^{2^2}+1, \quad 2^{2^3}+1, \quad 2^{2^4}+1, \quad 2^{2^5}+1, \quad \ldots,$$

that is, the numbers

$$3, \quad 5, \quad 17, \quad 257, \quad 65537, \quad 4294967297, \quad \ldots,$$

are all prime. The 1st five of these numbers are indeed prime, as may be verified with the help of a hand-calculator, but the sixth one is not – it is divisible by the prime number 641. What led Fermat to make the claim is unclear, but it is an example of how one can be misled by a pattern that holds only for the 1st few members of a sequence. Fermat's claim has proved to be rather ill-fated: after the 1st five numbers, not even one prime number has been discovered in the Fermat sequence! (In fairness, it must be added here that Fermat was not misled in this fashion very often – he was a mathematician with an uncanny instinct for spotting genuine patterns.)

★ ★ ★

3.1.6 There are more patterns waiting to be discovered. Earlier, we calculated the successive difference sequences for $\{n^3\}$, $\{n^4\}$ and $\{n^5\}$, till we finally reached a constant sequence – a sequence of the form $\{a, a, a, \ldots\}$, for some constant a. In the case of $\{n^2\}$, the constant 'a' was found to be 2.

What is the constant for $\{n\}$? This can be computed quickly. We construct the array for $\{n\}$:

0		1		2		3		4		5		6
	1		1		1		1		1		1	

The constant for $\{n\}$ is seen to be 1.

We now prepare a table of results as follows:

Sequence	$\{n\}$	$\{n^2\}$	$\{n^3\}$	$\{n^4\}$	$\{n^5\}$
Constant	1	2			

(Please fill up the blank slots.) Do you see a pattern in the last row? – there *is* a pattern, and it is a very pretty one! On the basis of the pattern, what do you expect the constants to be for the sequences $\{n^6\}$, $\{n^7\}$, $\{n^8\}$ and $\{n^9\}$?

3.2 A symbolic apparatus

It is appropriate at this stage to pause and re-examine whatever we have discovered. Specifically, we ask the following question: do we really 'understand' what is going on? Just as a scientist needs to pause at the conclusion of his experimentation and see if his findings make sense and 'fit together', similarly, we need to check whether our findings make sense. This section is devoted to doing such a check.

We start by introducing a few standard symbols. Let F be a given sequence whose n^{th} term is $F(n)$. We denote the difference sequence of F by $D(F)$; thus, the nth term of $D(F)$ is $F(n+1) - F(n)$. For example, if F is $\{n\}$:

$$F = \Big\langle 0,\ 1,\ 2,\ 3,\ 4,\ \dots \Big\rangle,$$

then

$$D(F) = \Big\langle 1,\ 1,\ 1,\ 1,\ 1,\ \dots \Big\rangle,$$

that is,

$$D(\{n\}) = \{1\}.$$

This result may be deduced symbolically as follows: the nth term of $\{n\}$ is n, so the nth term of $D(\{n\})$ is $(n+1) - n = 1$.

What does $D(\{n^2\})$ equal? The nth term of $\{n^2\}$ is n^2, so the nth term of $D(\{n^2\})$ is

$$(n+1)^2 - n^2 = 2n + 1,$$

which means that we have the relation

$$D\left(\{n^2\}\right) = \{2n+1\}.$$

Next, for the sequence whose n^{th} term is $2n+1$, the $(n+1)^{\text{th}}$ term is $2(n+1)+1 = 2n+3$. It follows that the nth term of $D(\{2n+1\})$ is

$$(2n+3) - (2n+1) = 2$$

and so $D(\{2n+1\}) = \{2\}$, that is, $D(\{2n+1\})$ is a sequence of 2s. Note that this can be written as

$$D_2\left(\{n^2\}\right) = \{2\},$$

where 'D_2' refers to the operation of taking differences twice over. ('D_3' would refer to the operation of taking differences *thrice* over, while 'D_1' is the same as 'D'.)

Notice that merely by playing around with the symbols, we have found something which earlier had been found by "number crunching". This shows the power of the symbolic method. Can we do the same for $\{n^3\}$?

The nth term of the sequence now is n^3. The nth term of its difference sequence is, therefore,

$$(n+1)^3 - n^3 = 3n^2 + 3n + 1$$

(via the expansion $(n+1)^3 = n^3 + 3n^2 + 3n + 1$). We write this in the form

$$D\left(\{n^3\}\right) = \{3n^2 + 3n + 1\}.$$

Next, the $(n+1)^{\text{th}}$ term of $\{3n^2 + 3n + 1\}$ is

$$3(n+1)^2 + 3(n+1) + 1 = 3n^2 + 9n + 4,$$

so the nth term of $D\left(\{3n^2 + 3n + 1\}\right)$ is

$$\left(3n^2 + 9n + 4\right) - \left(3n^2 + 3n + 1\right) = 6n + 3.$$

We write this result as

$$D_2\left(\{n^3\}\right) = \{6n + 3\}.$$

Finally, the nth term of $D\left(\{6n+3\}\right)$ is $(6(n+1)+3) - (6n+3) = 6$, so

$$D_3\left(\{n^3\}\right) = \{6\},$$

that is, the sequence of 3rd differences of $\{n^3\}$ is a sequence of 6s. This tallies exactly with the earlier results!

Exercises

3.2.1 Show that

$$\begin{aligned} D_1\left(\{n^4\}\right) &= \{4n^3 + 6n^2 + 4n + 1\}, \\ D_2\left(\{n^4\}\right) &= \{12n^2 + 24n + 14\}, \\ D_3\left(\{n^4\}\right) &= \{24n + 36\}, \\ D_4\left(\{n^4\}\right) &= \{24\}. \end{aligned}$$

3.2.2 Let $A = \{n^5\}$. Compute the sequences $D_1(A)$, $D_2(A)$, ..., down to $D_5(A)$.

3.2.3 We have already seen that

$$\begin{aligned} D\left(\{n^2\}\right) &= \{2n + 1\}, \\ D\left(\{n^3\}\right) &= \{3n^2 + 3n + 1\}. \end{aligned}$$

Note that in both cases, the degree of the expression on the right side is 1 less than the degree of the expression on the left side. Is this always so? The answer is yes, but to prove this we need more powerful machinery, namely the result that if k is any positive integer and n any number, then

$$(n+1)^k - n^k = kn^{k-1} + \frac{k(k-1)}{2}n^{k-2} + \cdots + 1.$$

For example, with $k = 5$, we have

$$(n+1)^5 - n^5 = 5n^4 + 10n^3 + 10n^2 + 5n + 1.$$

(This comes from the binomial theorem.) Use the theorem to show the following: *If F is generated by a polynomial of degree k, then $D(F)$ is generated by a polynomial of degree $k-1$.*

3.2.3 Let A and B be two sequences. Define the sequence $A+B$ to be the one whose n^{th} term is $A(n) + B(n)$; likewise, define the sequence $A-B$ to be the one whose n^{th} term is $A(n) - B(n)$. Show that

$$D(A+B) = D(A) + D(B),$$
$$D(A-B) = D(A) - D(B).$$

3.2.4 Let A and B be two sequences. Define the sequence AB to be the sequence whose n^{th} term is $A(n)B(n)$. Find $D(AB)$ in terms of $D(A)$ and $D(B)$.

Chapter 4

The Generating Formula

4.1 Introduction

Often, when mathematicians do research they come up with number sequences whose generating formula is unknown. One approach used by them in such cases is the method of differences—the steps outlined in earlier chapters. That is,

- we examine the sequence of 1st differences for patterns;
- if the pattern is not revealed, then we proceed to the sequence of 2nd differences;
- if the pattern is still not revealed, then we examine the sequence of 3rd differences;

and so on. Sometimes, if the sequence is not too "badly behaved", the method works: at some stage we reach a sequence which is either familiar or in some way easy to deal with, and after that the investigation becomes routine.

The method could be described as one of "filtration", because what is done at each stage is to filter away the "leading" component of the sequence, and this is repeated till the generating formula becomes clear. Another name that could be given to the method is "spectral analysis", because what we are doing is very similar to what a prism does to a beam of sunlight—it splits it up into its individual components.

The idea is best grasped through an example. Consider the sequence $A = \{2n^2 + 3n\}$:

$$A = \Big\langle\, 0,\ 5,\ 14,\ 27,\ 44,\ 65,\ 90,\ \ldots\, \Big\rangle.$$

Its array of differences is shown below.

0		5		14		27		44		65		90
	5		9		13		17		21		25	
		4		4		4		4		4		4

Observe that the sequence of 2nd differences is constant, and also that the constant is 4, which is *twice* the constant for $\{n^2\}$. (Earlier we had found that the constant for $\{n^2\}$ was 2.) This is not surprising, as the leading term in the generating formula of A is $2n^2$. Now let us "peel" or "filter" away the $2n^2$ portion of the sequence by doing term-by-term subtraction. That is, we compute the sequence $C = A - B$, where $B = \{2n^2\}$:

$$\begin{aligned} A &= \left\langle 0,\ 5,\ 14,\ 27,\ 44,\ 65,\ 90,\ \ldots \right\rangle, \\ B &= \left\langle 0,\ 2,\ 8,\ 18,\ 32,\ 50,\ 72,\ \ldots \right\rangle, \\ \therefore\ C &= \left\langle 0,\ 3,\ 6,\ 9,\ 12,\ 15,\ 18,\ \ldots \right\rangle. \end{aligned}$$

Having done the "filtering", the generating formula for the "residue" C is easy to see: C is just $\{3n\}$. (The first term is 3, the 2nd term is $2 \times 3 = 6$, the 3rd term is $3 \times 3 = 9$, and so on.)

Observe exactly what we have done. Through the difference array of the sequence, we were able to identify the leading term in its generating formula (in the above example, the leading term was $2n^2$). After filtering out this term we obtained a residue, a simpler sequence. A study of the difference array of this sequence now revealed *its* leading term, and so on. *This is the "method of differences".*

4.2 A worked example

We illustrate the method through a worked example. Consider the sequence $F = \{n^3 + 2n\}$:

$$F = \left\langle 0,\ 3,\ 12,\ 33,\ 72,\ 135,\ 228,\ 357,\ 528,\ \ldots \right\rangle.$$

Now, suppose that we have only this sequence to work with; *we do not have its generating formula*. However, we have available for study as many terms of F as needed. Is there some manner by which we can find the generating formula? We approach the problem by examining the successive difference sequences. The array of difference sequences is displayed below.

0		3		12		33		72		135		228
	3		9		21		39		63		93	
		6		12		18		24		30		
			6		6		6		6			

Note that the pattern in the sequence of 1st differences is not clear (is there a pattern at all?); so we go on to the 2nd level. Things now start to look more familiar; we obtain the sequence shown below:

$$D_2(F) = \left\langle 6,\ 12,\ 18,\ 24,\ 30,\ \ldots \right\rangle,$$

where each term appears to be greater than the one which precedes it by the same amount, namely 6. Because of this, the sequence of 3rd differences is a constant sequence, the constant being 6. Haven't we seen this happen somewhere earlier? Yes! – it happens in the case of the sequence $\{n^3\}$. For that sequence too the constant in the 3rd level is 6, so it seems reasonable to assume that the leading role in the generating formula is played by a *cubic* term, that is, n^3, and not, say, n^4 or n^5 or a higher (or lower) power of n. The constant ($= 6$) is the same as that for $\{n^3\}$, so we do a term-by-term subtraction of $\{n^3\}$ from F. Now,

$$\begin{aligned} \{n^3\} &= \left\langle 0,\ 1,\ 8,\ 27,\ 64,\ 125,\ 216,\ 343,\ \ldots \right\rangle, \\ F &= \left\langle 0,\ 3,\ 12,\ 33,\ 72,\ 135,\ 228,\ 357,\ \ldots \right\rangle, \end{aligned}$$

so the sequence $G = F - \{n^3\}$ is

$$G = \left\langle 0,\ 2,\ 4,\ 6,\ 8,\ 10,\ 12,\ 14,\ \ldots \right\rangle.$$

This is a nice-looking sequence and we should be able to guess its generating formula immediately. It appears to be just $\{2n\}$, the sequence of even numbers. But if this is so, then F must be $\{n^3 + 2n\}$. This is exactly the formula with which we had started! We should feel happy in our achievement: we have succeeded in recovering the generating formula of a sequence merely by examining the array of differences.

⋆ ⋆ ⋆

Here is another such example. Suppose that the sequence given to us is H, where

$$H = \left\langle 0,\ 6,\ 34,\ 96,\ 204,\ 370,\ 606,\ 924,\ 1336,\ \ldots \right\rangle.$$

The array of differences of H is shown below.

0		6		34		96		204		370
	6		28		62		108		166	
		22		34		46		58		
			12		12		12			

Once again, the pattern in the sequence of 1st differences is not readily seen, so we proceed to the 2nd level and then to the 3rd level. Here we reach a constant sequence; the sequence of 3rd differences is $\{12\}$, so it seems likely that the leading term in the generating formula is a cubic one. Moreover, as the constant $(= 12)$ is 2×6 or twice the constant for $\{n^3\}$, the leading term must be $2n^3$ and not n^3. We therefore subtract $\{2n^3\}$ from the given sequence, that is, we subtract

$$\left\langle 0,\ 2,\ 16,\ 54,\ 128,\ 250,\ 432,\ 686,\ \ldots \right\rangle$$

from

$$\left\langle 0,\ 6,\ 34,\ 96,\ 204,\ 370,\ 606,\ 924,\ \ldots \right\rangle$$

and obtain

$$\left\langle 0,\ 4,\ 18,\ 42,\ 76,\ 120,\ 174,\ 238,\ \ldots \right\rangle.$$

For this sequence, too, the pattern (is there one?) is not readily seen, so we repeat the procedure—we construct *its* array of differences:

4		18		42		76		120		174
	14		24		34		44		54	
		10		10		10		10		

For this array, the sequence of 2nd differences is a constant sequence. What does this suggest? Clearly that the generating formula involves the term n^2. The constant reached is 10, and since 10 is 5×2 or five times the constant for $\{n^2\}$, the starting sequence ought to involve the term $5n^2$. Accordingly, that is we now subtract $\{5n^2\}$ from the latest sequence, i.e., we subtract

$$\langle 0,\ 5,\ 20,\ 45,\ 80,\ 125,\ 180,\ 245,\ \ldots \rangle$$

from

$$\langle 0,\ 4,\ 18,\ 42,\ 76,\ 120,\ 174,\ 238,\ \ldots \rangle.$$

The result is

$$\langle 0,\ -1,\ -2,\ -3,\ -4,\ -5,\ -6,\ -7,\ \ldots \rangle,$$

which surely is just $\{-n\}$! Therefore, it seems that the original sequence must have been

$$\{2n^3 + 5n^2 - n\},$$

and this certainly fits the facts; for example, for $n = 5$, the formula gives $(2 \times 5^3) + (5 \times 5^2) - 5 = 370$, which is correct (please examine the list given earlier). So, we have found the answer! Or have we?

4.3 Warning!

The reader must be warned of a very real danger at this stage. Suppose that we are given a sequence whose 1st ten terms fit into a certain pattern. Does this imply that the terms to follow (the $11^{\text{th}}, 12^{\text{th}}, 13^{\text{th}}, \ldots$ terms) also follow the same pattern? Not at all – it may happen that the observed pattern is a spurious one and that the real pattern is much more subtle and can be seen only after examining a larger number of terms. It is very important to keep this in mind, because many serious errors have been made in precisely this manner, even by famous mathematicians and scientists (we have already mentioned one such instance, attributed to Fermat). What we mean is this: in the 2nd of the two examples presented above, when we claim that the sequence

$$\langle 0,\ -1,\ -2,\ -3,\ -4,\ -5,\ -6,\ -7,\ \ldots \rangle$$

is $\{-n\}$, the claim has been made *under the assumption that the visible pattern continues as one expects it to, in the "obvious" way.* Strictly speaking, if we are given only the initial portion of the sequence, say only the numbers

$$0,\ 6,\ 34,\ 96,\ 204,\ 370,\ 606,\ 924,\ 1336,\ 1854,\ \ldots,$$

(the '...' tells us that there are more terms to follow, only we do not know quite what they are), we *cannot* deduce that the next term after 1854 is 2490 (which is the value taken by the expression $2n^3 + 5n^2 - n$ when $n = 10$). All we can say with certainty is the following:

> *A formula has been found, namely* $2n^3 + 5n^2 - n$*, that fits the initial portion of the given sequence, and according to which the* 10th *term should be 2490. However, as we are not given any further data, there is no way whatever by which we may verify the prediction. Indeed, there may be other formulas that fit the same initial portion of the sequence and which give different values for the* 10th *term.*

Here is another such instance. Consider a sequence A, whose 1st three terms are as given below.

$$A = \left\langle 1,\ 2,\ 4,\ \ldots \right\rangle.$$

(We do not have very much on which to base a prediction!) The sequence can be generated by the formula 2^n (because $2^0 = 1$, $2^1 = 2$, $2^2 = 4$) and also by the formula

$$\frac{n^2 + n + 2}{2}.$$

To see why, note that $(0^2 + 0 + 2)/2 = 1$, $(1^2 + 1 + 2)/2 = 2$, and $(2^2 + 2 + 2)/2 = 4$.

The 1st formula predicts that the next term (corresponding to $n = 3$) is 8, while the 2nd formula gives a value of 7. Which is it, actually? Given only the 1st three terms, *we have no basis whatsoever for choosing between the two predictions*. If, at a later point in time, the next term were revealed to us to be, say, 8, then we could immediately discard the 2nd formula, but this still would not imply that the 1st formula is correct—because there may be yet other formulas that fit the sequence 1, 2, 4, 8, ... Indeed, here is such a formula:

$$\frac{n^3 + 5n + 6}{6}.$$

Please check that it does indeed fit! (For $n = 1$, 2 and 3 it gives 2, 4 and 8 respectively, as it should, and it predicts that the 4th term is 15.)

4.4 Interlude: Hypotheses in Science

If you have been thrown into confusion by these comments, you can take comfort in the fact that this type of uncertainty is a very basic aspect of science. It is not uncommon in science for hypotheses to be put forward *purely* on the basis of an observed pattern. Subsequent research may confirm or disprove the correctness of the hypothesized model, i.e., how well it fits the facts, and if it does fit, later researchers may try to find some theoretical justification behind it. We give two examples of this phenomenon below.

• *Ptolemy and planetary motion* Long ago, back in ancient Greece, the astronomer Ptolemy put forward a model to account for the observed movements of the stars and planets. The basic facts which the theory sought to explain were: (a) the observed motion of the planets, which seemed to move against the background of the stars; and (b) the curious phenomenon of retrograde motion, whereby some planets (Mars, Jupiter and Saturn) seem at certain moments to turn back in their orbits, move in the reverse direction for some time, and then turn back once more to continue in their original tracks. Ptolemy suggested that the planets move in epicycles. The Greeks loved circles, regarding it one of the perfect forms, and wanted a theory which would incorporate circles into the scheme of things; and Ptolemy's model did just this. The theory certainly matched observation; but only for a while! Eventually it was found that Ptolemy's "formula" failed to account for many phenomena. In a desperate effort to repair the defects, later astronomers added more and more epicycles, until the Solar System seemed full of epicycles and little else. Ultimately the theory collapsed under its own weight, because another theory came along – that of Copernicus, who placed the Sun rather than the Earth at the centre of the Solar System. So, the "formula" put forward by Ptolemy had to be discarded and a new one took its place.

• *Alfred Wegener and the movements of the continents* In the early decades of the 20th century the German meteorologist, Alfred Wegener, suggested that the way in which the continents of Africa and South America seem to fit together, coastline upon coastline, is not a coincidence. He obtained data on the shapes of the continental shelves adjoining the two continents and found that there was a fit even along these margins. This strengthened his conviction, and he put forward the hypothesis that, at some point in time, the two continents must have been together as one landmass. He was bold enough to publish his theory, which had no backing whatsoever except for the observed geometric fit. As he was a meteorologist rather than a geologist, his ideas were met with considerable skepticism. The main stumbling block was, of course, the absence of a theory that could explain how such massive objects like continents could move upon the surface of the Earth. Gradually, however, evidence gathered in favour of the theory—research on the mid-Atlantic ridge, fossil evidence, evidence based on the magnetic alignment of the rocks in the Atlantic Ocean, and so on. Finally, the theory was accepted, after a mechanism that could move continents was found. Today the phenomenon is well known to all of us; we call it "continental drift". It is worth realizing, however, that Wegener well and truly stuck his neck out in putting forward the theory—on the basis only of a geometric pattern. (This may be regarded as the analogue of guessing a generating formula based only on the initial portion of a sequence.)

⋆ ⋆ ⋆

The moral should be very clear: given only the initial portion of a sequence, many formulas can be found that generate the given numbers and, without further data, we have no logical basis for choosing between them. More generally, a pattern that holds for an initial portion of a sequence, *however long*, does not necessarily hold for the entire sequence. We have already seen examples of this. Here is another curious example, provided by the sequence $\{n^2+n+41\}$. Its 1st 40 entries, for $n = 0, 1, 2, \ldots, 39$, are given below (the table is to be read row-wise: the entry 41 corresponds to $n = 0$; 43 to $n = 1$; 911 to $n = 29$, 971 to $n = 30$; 1601 to $n = 39$, and so on):

41,	43,	47,	53,	61,
71,	83,	97,	113,	131,
151,	173,	197,	223,	251,
281,	313,	347,	383,	421,
461,	503,	547,	593,	641,
691,	743,	797,	853,	911,
971,	1033,	1097,	1163,	1231,
301,	1373,	1447,	1523,	1601.

It turns out (surprise!) that these 40 numbers are *all prime*. Sadly, however, the very next value, taken when $n = 40$, is $40^2+40+41 = 1681$, which is not prime. (Proof: $1681 = 41^2$.)

Exercises

In the exercises below, you are asked to find the generating formulas of the given sequences using the method of differences. Keeping in mind the comments made above, you may freely assume that the patterns uncovered do indeed persist! – after all, it is being done only for the sake of an exercise.

4.4.1 Find formulas that fit the following sequences:

$$\text{(i)} \quad \langle 1,\ 3,\ 7,\ 13,\ 21,\ 31,\ 43,\ 57,\ 73,\ 91,\ \ldots \rangle;$$

$$\text{(ii)} \quad \langle 3,\ 8,\ 19,\ 36,\ 59,\ 88,\ 123,\ \ldots \rangle;$$

$$\text{(iii)} \quad \langle 0,\ 9,\ 30,\ 75,\ 156,\ 285,\ 474,\ 735,\ \ldots \rangle;$$

$$\text{(iv)} \quad \langle 0,\ 2,\ 24,\ 108,\ 320,\ 750,\ 1512,\ 2744,\ \ldots \rangle.$$

4.4.2 Find two formulas that fit the sequence P, where

$$P = \langle 1,\ 3,\ 9,\ \ldots \rangle.$$

⋆ ⋆ ⋆

We now illustrate the working of the method of differences via three more problems, all geometrical in nature.

4.5 Triangular numbers

Problem Define the triangular numbers as shown in the arrays of dots in Figure A.4.1, then find a formula (in terms of n) for the nth triangular number T_n.

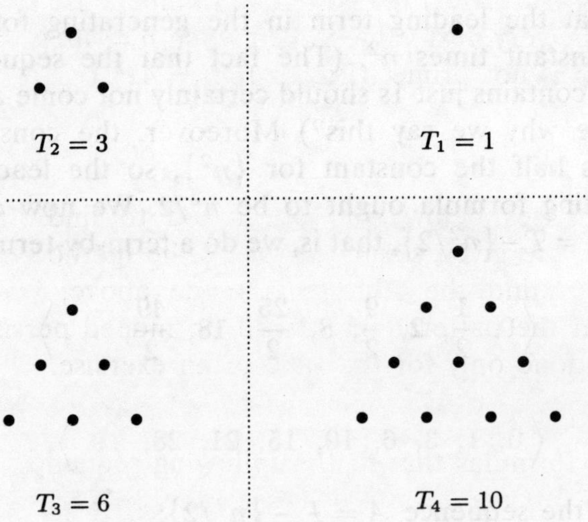

Figure A.4.1 *Triangular numbers*

The reason for the name 'triangular number' should be clear: the dot arrays shown are all triangular in shape. Write T_n for the number of dots in the nth such array. Then

$$T_1 = 1, \quad T_2 = 3, \quad T_3 = 6, \quad T_4 = 10, \quad T_5 = 15, \quad \ldots .$$

Obviously, a triangular shape can be constructed using dots only if the number of dots is a triangular number. It should be fairly clear that

$$T_n = 0 + 1 + 2 + 3 + \cdots + (n-1) + n,$$

for each new row of dots contains one more dot than the previous row. So, by doing a few more calculations, we obtain more members of the sequence T. Here is the sequence:

$$T = \Big\langle\, 0,\ 1,\ 3,\ 6,\ 10,\ 15,\ 21,\ 28,\ 36,\ 45,\ 55,\ \ldots \Big\rangle,$$

Our task now is to find a formula for T_n, the nth triangular number. We apply the method of differences to T and obtain the following array:

0		1		3		6		10		15		21		28
	1		2		3		4		5		6		7	
		1		1		1		1		1		1		

The sequence of 2nd differences consists only of 1s, so we may suppose that the leading term in the generating formula for T is some constant times n^2. (The fact that the sequence of 2nd differences contains just 1s should certainly not come as a surprise. Do you see why we say this?) Moreover, the constant for the sequence is half the constant for $\{n^2\}$, so the leading term in the generating formula ought to be $n^2/2$. We now compute the sequence $A = T - \{n^2/2\}$, that is, we do a term-by-term subtraction of

$$\left\langle 0,\ \frac{1}{2},\ 2,\ \frac{9}{2},\ 8,\ \frac{25}{2},\ 18,\ \frac{49}{2},\ \ldots \right\rangle$$

from T:

$$\left\langle 0,\ 1,\ 3,\ 6,\ 10,\ 15,\ 21,\ 28,\ \ldots \right\rangle,$$

and obtain the sequence $A = T - \{n^2/2\}$:

$$A = \left\langle 0,\ \frac{1}{2},\ 1,\ \frac{3}{2},\ 2,\ \frac{5}{2},\ 3,\ \frac{7}{2},\ \ldots \right\rangle.$$

This sequence can be written in a nicer form by expressing all the number as fractions with a denominator of 2:

$$A = \left\langle \frac{0}{2},\ \frac{1}{2},\ \frac{2}{2},\ \frac{3}{2},\ \frac{4}{2},\ \frac{5}{2},\ \frac{6}{2},\ \frac{7}{2},\ \ldots \right\rangle,$$

and it appears from this that A is just $\{n/2\}$. Therefore, the sequence initially given must have been

$$\left\{ \frac{n^2}{2} + \frac{n}{2} \right\},$$

i.e., $\{n(n+1)/2\}$. The conclusion is thus,

$$T_n = 0 + 1 + 2 + \cdots + n = \frac{n(n+1)}{2}.$$

That is, the nth triangular number is $n(n+1)/2$. The formula is correct! It is easily verified for particular values of n; for instance, for $n = 10$, the formula states that

$$T_{10} = 0 + 1 + 2 + 3 + \cdots + 10 = 55,$$

which is true.

Even better, there is a nice way to *prove* the formula; but we shall not spoil your fun by giving away the proof. Please find it on your own. (Hint: Write S, as the sum defined by

$$S = 1 + 2 + 3 + \cdots + (n-1) + n.$$

Observe that the sum can also be written as

$$S = n + (n-1) + (n-2) + \cdots + 2 + 1.$$

Now, add the two expressions for S, term by term. What do you find? Does this yield a formula for S?)

4.6 Diagonals of a polygon

Problem Find a formula, in terms of n, for the number of diagonals of an n-sided convex polygon ('n-gon' for short).

Let d_n denote the number of diagonals of the n-gon. Since a polygon cannot have less than 3 sides, we have $n \geq 3$. A triangle has no diagonals, so $d_3 = 0$. A quadrilateral has 2 diagonals, so $d_4 = 2$, and a pentagon has 5 diagonals, so $d_5 = 5$.

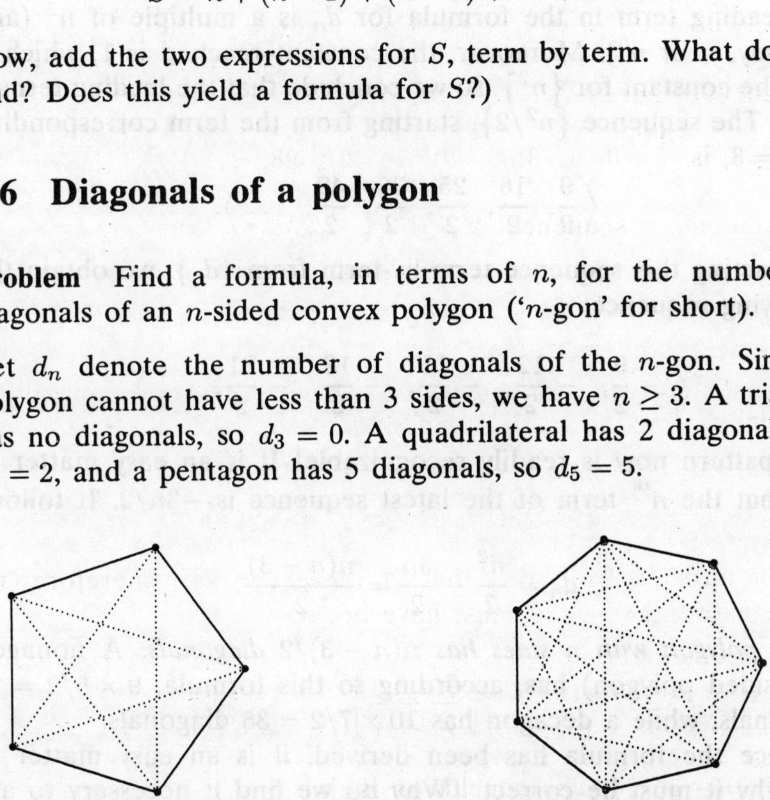

Pentagon
(5 diagonals)

Heptagon
(14 diagonals)

Figure A.4.2 *How many diagonals does a polygon have?*

Continuing in this way, we build the following table:

n	3	4	5	6	7	8
d_n	0	2	5	9	14	20

We are now ready to apply the method of differences. The array of differences is displayed below.

0		2		5		9		14		20
	2		3		4		5		6	
		1		1		1		1		

As in Problem 1, the 2nd differences are all 1, so we conclude that the leading term in the formula for d_n is a multiple of n^2 (and not, say, n or n^3). Moreover, the constant reached is 1, which is half the constant for $\{n^2\}$, so we conclude that the leading term is $n^2/2$. The sequence $\{n^2/2\}$, starting from the term corresponding to $n = 3$, is

$$\left\langle \frac{9}{2}, \frac{16}{2}, \frac{25}{2}, \frac{36}{2}, \frac{49}{2}, \ldots \right\rangle.$$

Subtracting this sequence term-by-term from $\{d_n\}$, we obtain the following sequence:

$$\left\langle -\frac{9}{2}, -\frac{12}{2}, -\frac{15}{2}, -\frac{18}{2}, -\frac{21}{2}, \ldots \right\rangle.$$

The pattern now is readily recognizable! It is an easy matter to see that the nth term of the latest sequence is $-3n/2$. It follows that

$$d_n = \frac{n^2}{2} - \frac{3n}{2} = \frac{n(n-3)}{2}.$$

So, a polygon with n sides has $n(n-3)/2$ diagonals. A nonagon (a 9-sided polygon) has, according to this formula, $9 \times 6/2 = 27$ diagonals, while a decagon has $10 \times 7/2 = 35$ diagonals.

Once the formula has been derived, it is an easy matter to see why it must be correct. (Why do we find it necessary to ask whether the formula is correct? Remember that the formula has been found only on the basis of an observed pattern. How do we know for sure whether the pattern will continue?) Here are the details.

Since each diagonal has two end-points, each diagonal corresponds to a *pair* of vertices of the polygon. How many such pairs of vertices are there in all? As there are n vertices, and each of the n vertices can be paired with every other, the number of pairs of vertices possible ought to be $n \times (n-1)$. However, in this count, each pair has been accounted for *twice*. For example, the diagonal AC would be accounted for as $\{A, C\}$ as well as $\{C, A\}$. So we must *halve* the expression which we have obtained. Accordingly, we see that the number of pairs of vertices is

$$\frac{n \times (n-1)}{2}.$$

Do all these pairs correspond to diagonals of the polygon? Not so!—for the sides of the polygon also correspond to pairs of vertices. How many sides are there?—clearly n. This must be subtracted from the expression shown above. It follows that the number of diagonals is $d_n = \frac{1}{2}n(n-1) - n$, that is,

$$d_n = \frac{n^2 - 3n}{2},$$

which is just what we found earlier on the basis of the observed pattern.

★ ★ ★

Problems of this kind, which involve some sort of "counting", offer a rich source of problems. We shall return to this theme in a later part (Section 6.1) of the book.

4.7 Pentagonal numbers

Problem Define the pentagonal numbers as shown in the arrays of dots in Figure A.4.3, then find a formula (in terms of n) for the nth pentagonal number P_n.

The reason for calling the numbers "pentagonal" is easy to see: the dots form a lattice-work of pentagons, the distances between neighbouring dots being always equal to 1. (The diagrams are not all drawn to the same scale.)

We observe that $P_1 = 1$, $P_2 = 5$, $P_3 = 12$, $P_4 = 22$, $P_5 = 35$, $P_6 = 51$, ... Denoting the sequence by P, we have

$$P = \left\langle\, 1,\ 5,\ 12,\ 22,\ 35,\ 51,\ \ldots \right\rangle.$$

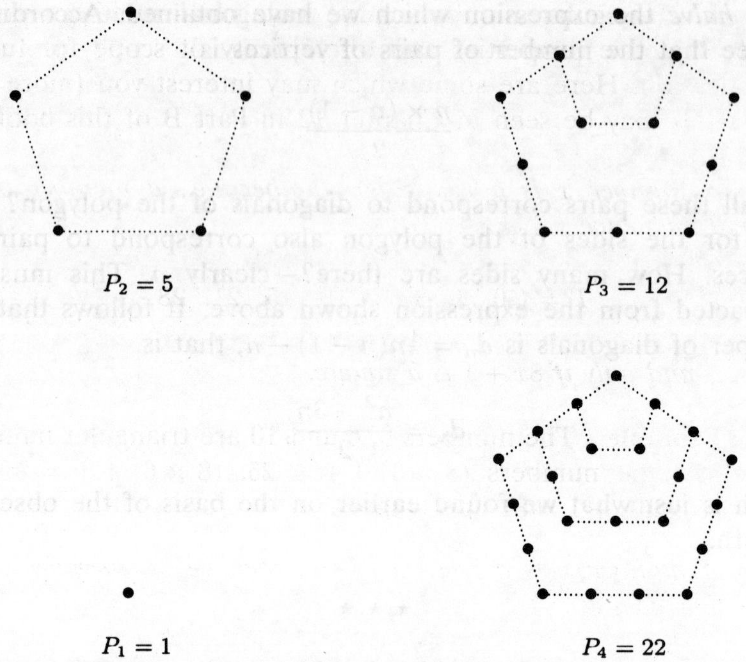

Figure A.4.3 *The first four pentagonal numbers*

To find a formula for P_n we compute its array of differences. The result is displayed below.

1		5		12		22		35		51
	4		7		10		13		16	
		3		3		3		3		3

A constant is reached in the 2nd differences, and the constant is 3, which is 3/2 times the constant for $\{n^2\}$; so we subtract $\{3n^2/2\}$ from P. We obtain the sequence

$$P - \left\{\frac{3n^2}{2}\right\} = \left\langle -\frac{1}{2}, -\frac{2}{2}, -\frac{3}{2}, -\frac{4}{2}, -\frac{5}{2}, \ldots \right\rangle.$$

The nth term of the last sequence is "clearly" seen to be $-n/2$. It follows that

$$P_n = \frac{3n^2}{2} - \frac{n}{2} = \frac{n(3n-1)}{2}.$$

So the nth pentagonal number is $n(3n-1)/2$.

★ ★ ★

Remark Triangular numbers and pentagonal numbers possess many attractive properties and offer plenty of scope for further investigation. Here are some which may interest you (more such properties may be seen in Chapter 12 in Part B of this book).

- *A number x is a triangular number, i.e., a member of the T-sequence*

  $$T = \left\langle 0,\ 1,\ 3,\ 6,\ 10,\ 15,\ 21,\ 28,\ \dots \right\rangle$$

  if and only if $8x + 1$ is a square.

  Example The numbers 3, 6 and 10 are triangular numbers, and the numbers $(8 \times 3) + 1 = 25$, $(8 \times 6) + 1 = 49$ and $(8 \times 10) + 1 = 81$ are squares.

- *A number x is a pentagonal number, i.e., a member of the P-sequence,*

  $$P = \left\langle 0,\ 1,\ 5,\ 12,\ 22,\ 35,\ 51,\ \dots \right\rangle$$

  if and only if $24x + 1$ is a square.

  Example The numbers 1, 5 and 12 are pentagonal numbers, and the numbers $(24 \times 1) + 1 = 25$, $(24 \times 5) + 1 = 121$ and $(24 \times 12) + 1 = 289$ are squares.

- *Every positive integer can be written as a sum of three triangular numbers.*

  Example Consider the numbers 40, 50 and 60. We have $40 = 15 + 15 + 10$, $50 = 28 + 21 + 1$ and $60 = 36 + 21 + 3$.

The 1st two statements are easy to prove, but the 3rd poses a considerable challenge. The reader is referred to books on number theory for further study.

4.8 Sums of squares

Encouraged by our successes, we go on to ask for a nice formula for the sum

$$0^2 + 1^2 + 2^2 + 3^2 + \cdots + n^2,$$

Therefore,

$$A = \frac{n^3}{3} + \frac{n^2}{2} + \frac{n}{6}.$$

This is exactly the formula obtained earlier!

Exercises

Listed here are a few exercises to be done along the lines of the examples discussed above. In each case, n represents an arbitrary whole number.

4.9.1 Find a formula for the sum

$$0^3 + 1^3 + 2^3 + \cdots + (n-1)^3 + n^3.$$

4.9.2 Find a formula for the sum

$$0^4 + 1^4 + 2^4 + \cdots + (n-1)^4 + n^4.$$

4.9.3 Find a formula for the sum of the squares of the 1st n odd numbers, that is, a formula for the nth term of the sequence

$$\left\langle 1^2,\ 1^2 + 3^2,\ 1^2 + 3^2 + 5^2,\ 1^2 + 3^2 + 5^2 + 7^2,\ \ldots \right\rangle.$$

4.9.4 Find a formula for the nth term of the sequence

$$\left\langle 1^2,\ 1^2 + 4^2,\ 1^2 + 4^2 + 7^2,\ 1^2 + 4^2 + 7^2 + 10^2,\ \ldots \right\rangle.$$

Chapter 5

Exponential Sequences

5.1 Sequences of the form $\{a^n\}$

Having studied in some detail the sequences that arise from polynomial functions, we now turn to other types of sequences; the most familiar one is the sequence of powers of some fixed number. (This is a particular case of an exponential sequence.) We quickly encounter unexpected behaviour.

5.1.1 The Sequence $\{(-1)^n\}$

Consider the sequence of alternating 1s and -1s, $\{(-1)^n\}$:

$$\Big\langle 1,\ -1,\ 1,\ -1,\ 1,\ -1,\ 1,\ -1,\ \ldots \Big\rangle.$$

Its table of differences is shown below.

1		−1		1		−1		1		−1
	−2		2		−2		2		−2	
		4		−4		4		−4		
			−8		8		−8			

How curious! – it appears that we shall never reach a constant sequence; we seem to get ever larger numbers! Note the unexpected presence of the powers of 2, that too with alternating signs. How do you explain the appearance of these numbers?

5.1.2 The Sequence $\{2^n\}$

Next, we work with $\{2^n\}$, the sequence of powers of 2. The result is displayed below.

1		2		4		8		16		32		64
	1		2		4		8		16		32	
		1		2		4		8		16		
			1		2		4		8			

that is, a formula for the sum of the 1st $n+1$ square numbers. The 1st few sums, corresponding to $n = 0, 1, 2, 3, \ldots$ are

$$0,\ 0+1=1,\ 0+1=1,\ 0+1+4=5,\ 0+1+4+9=14,$$

and we obtain a sequence A, given by

$$A = \left\langle 0,\ 1,\ 5,\ 14,\ 30,\ 55,\ 91,\ 130,\ 194,\ 275,\ \ldots \right\rangle,$$

whose n^{th} term is the sum $0^2 + 1^2 + 2^2 + \cdots + n^2$. We now use the method of differences to uncover the generating formula of A. Here is the array of differences of A:

0		1		5		14		30		55		91
	1		4		9		16		25		36	
		3		5		7		9		11		
			2		2		2		2			

The 3rd differences are all equal to 2, which is 1/3 the constant for $\{n^3\}$. Conclusion: The leading term in the generating formula for A must be 1/3 of n^3. So we compute the sequence $B = A - \{n^3/3\}$, i.e., we subtract

$$\left\langle 0,\ \frac{1}{3},\ \frac{8}{3},\ \frac{27}{3},\ \frac{64}{3},\ \frac{125}{3},\ \frac{216}{3},\ \ldots \right\rangle$$

from

$$\left\langle 0,\ 1,\ 5,\ 14,\ 30,\ 55,\ 91,\ \ldots \right\rangle$$

and obtain

$$B = \left\langle 0,\ \frac{2}{3},\ \frac{7}{3},\ \frac{15}{3},\ \frac{26}{3},\ \frac{40}{3},\ \frac{57}{3},\ \ldots \right\rangle$$

Since the generating formula for B is not obvious, we continue the process and obtain the array of differences for B:

$\frac{0}{3}$		$\frac{2}{3}$		$\frac{7}{3}$		$\frac{15}{3}$		$\frac{26}{3}$		$\frac{40}{3}$		$\frac{57}{3}$
	$\frac{2}{3}$		$\frac{5}{3}$		$\frac{8}{3}$		$\frac{11}{3}$		$\frac{14}{3}$		$\frac{17}{3}$	
		$\frac{3}{3}$		$\frac{3}{3}$		$\frac{3}{3}$		$\frac{3}{3}$		$\frac{3}{3}$		

The 2nd differences are all equal to 3/3 or 1, which implies a constant multiple of n^2. Moreover, the constant term is half the constant for $\{n^2\}$. So we subtract $\{n^2/2\}$ from B. That is, we subtract

$$\left\langle 0,\ \frac{1}{2},\ \frac{4}{2},\ \frac{9}{2},\ \frac{16}{2},\ \frac{25}{2},\ \frac{36}{2},\ \cdots \right\rangle$$

from

$$\left\langle 0,\ \frac{2}{3},\ \frac{7}{3},\ \frac{15}{3},\ \frac{26}{3},\ \frac{40}{3},\ \frac{57}{3},\ \cdots \right\rangle$$

and obtain a sequence $C = B - \{n^2/2\}$:

$$C = \left\langle 0,\ \frac{1}{6},\ \frac{2}{6},\ \frac{3}{6},\ \frac{4}{6},\ \frac{5}{6},\ \frac{6}{6},\ \cdots \right\rangle,$$

which is "seen" quite visibly to be $\{n/6\}$! (Of course, we deliberately wrote the sequence so that the fractions all have the same denominator.) Retracing our steps, we find that

$$A = \left\{ \frac{n^3}{3} + \frac{n^2}{2} + \frac{n}{6} \right\}.$$

This means that

$$0^2 + 1^2 + 2^2 + \cdots + n^2 = \frac{n^3}{3} + \frac{n^2}{2} + \frac{n}{6},$$

a remarkable formula! Let us check whether it is correct for some sample value of n, say $n = 10$. The sum on the left side is

$$0 + 1 + 4 + 9 + 16 + 25 + 36 + 49 + 64 + 81 + 100 = 385,$$

and the sum on the right side is

$$\frac{1000}{3} + \frac{100}{2} + \frac{10}{6} = \frac{2310}{6} = 385.$$

The values match—what a pleasing achievement!

A nicer way of writing the formula we have just found is

$$1^2 + 2^2 + \cdots + (n-1)^2 + n^2 = \frac{n(n+1)(2n+1)}{6}.$$

4.9 Proof of the formula

You may wonder whether there is a nice way of proving this formula, as there was for the formula $0+1+2+\cdots+n = n(n+1)/2$.

Indeed there is (in fact there are many nice proofs, but they are not so easy to find). One way is to use the identity

$$x^3 - (x-1)^3 = 3x^2 - 3x + 1$$

which is easily checked. In this identity, we substitute for x the values n, $n-1$, $n-2, \ldots, 2, 1$. Here is what we get.

$$\begin{aligned}
n^3 - (n-1)^3 &= 3n^2 - 3n + 1,\\
(n-1)^3 - (n-2)^3 &= 3(n-1)^2 - 3(n-1) + 1,\\
(n-2)^3 - (n-3)^3 &= 3(n-2)^2 - 3(n-2) + 1,\\
\cdots \quad \cdots &= \cdots \quad \cdots,\\
3^3 - 2^3 &= 3(3^2) - 3(3) + 1,\\
2^3 - 1^3 &= 3(2^2) - 3(2) + 1,\\
1^3 - 0^3 &= 3(1^2) - 3(1) + 1.
\end{aligned}$$

We now add all these resulting equations together that is, we total up the quantities on the left side and then we total up the quantities on the right side. A very convenient cancellation takes place on the left side–the *only* term that remains is n^3 (please check this!). By grouping together similar terms on the right side, we write it in the form $3A - 3B + C$, where

$$\begin{aligned}
A &= n^2 + (n-1)^2 + (n-2)^2 + \cdots + 2^2 + 1^2,\\
B &= n + (n-1) + (n-2) + \cdots + 2 + 1,\\
C &= 1 + 1 + 1 \cdots + 1 + 1 = n.
\end{aligned}$$

We already know the value of B:

$$B = n + (n-1) + (n-2) + \cdots + 1 = \frac{n(n+1)}{2}.$$

We therefore have the equation

$$n^3 = 3A - 3 \cdot \frac{n(n+1)}{2} + n,$$

from which it follows that

$$\begin{aligned}
3A &= n^3 + \frac{3n^2}{2} + \frac{3n}{2} - n\\
&= n^3 + \frac{3n^2}{2} + \frac{n}{2}.
\end{aligned}$$

This time, each sequence of differences is absolutely identical to its parent! It is clear that in this case too we shall never reach a sequence of constant differences.

We see here a fundamental difference between exponential sequences and polynomial sequences.

5.1.3 The Sequence $\{3^n\}$

Next, we experiment with $\{3^n\}$, the sequence of powers of 3. Here is its array of differences:

1		3		9		27		81		243
	2		6		18		54		162	
		4		12		36		108		
			8		24		72			
				16		48				
					32					

This time, each difference sequence is entirely different from each of its 'ancestors', and, as earlier, it looks as though a constant sequence will never be reached; indeed, the numbers seem to get progressively larger as we proceed downwards! Also, just as was observed for $\{(-1)^n\}$, there is the rather unexpected presence of the sequence $\{2^n\}$,

$$\left\langle 1,\ 2,\ 4,\ 8,\ 16,\ 32,\ \dots \right\rangle,$$

formed by the initial numbers of the different rows. How curious that in computing the difference array of $\{3^n\}$, we should meet $\{2^n\}$!

Note that in each horizontal row, each number is three times the number preceding it. For example, the 3rd row has the numbers

$$16,\ 48,\ 144,\ 432,\ \dots,$$

with $48 = 3 \times 16$, $144 = 3 \times 48$, and so on. This property is true of the original sequence, too, as its members are the successive powers of 3; so we have here a property that stays unchanged as we move from row to row. A similar statement can be made about the diagonals of the array which go from the top-left towards the bottom-right of the array: we find that each number is *twice* the preceding number. This can be seen for the diagonals $(1, 2, 4, \dots)$ and $(3, 6, 12, \dots)$. Here is another noteworthy observation: *No number appears more than once in the array.* A relevant question that may be asked here is: *Are these features genuine? Do they hold for the entire array?*

5.2 Mixed sequences

Next, we consider the case of mixed sequences, where the generating expression has polynomial as well as exponential terms. Consider the sequence $\{2^n + n\}$:

$$\langle 1,\ 3,\ 6,\ 11,\ 20,\ 37,\ 70,\ 135,\ \ldots \rangle.$$

Its array of differences is shown below:

1		3		6		11		20		37		70
	2		3		5		9		17		33	
		1		2		4		8		16		
			1		2		4		8			

Surprise!—by the 2nd level itself, the '$\{n\}$' has been completely erased from the original sequence, leaving behind only the '$\{2^n\}$'. Is this a general phenomenon? We experiment with $\{2^n + n^2\}$ to see what happens.

1		3		8		17		32		57		100
	2		5		9		15		25		43	
		3		4		6		10		18		
			1		2		4		8			

The '$\{n^2\}$' portion *has* indeed been erased, but by the 3rd level this time. The similarity of behaviour suggests that there is some general principle at work.

Exercises

5.2.1 Repeat the experiment with the sequences $\{4^n\}$ and $\{5^n\}$. What do you find? Keep a record of your findings—the list may prove useful at a later point in time.

5.2.2 Repeat the exercise on the following sequences:

(a) $\{3^n + n\}$

(b) $\{3^n + n^2\}$

(c) $\{3^n + n^2 + n\}$

(d) $\{2^n + n^3\}$

(e) $\{3^n + 2^n\}$

What features are to be found in the various difference arrays?

5.3 Seeking an explanation

The symbolic method discussed earlier now helps in understanding just what is going on. We consider the sequence $\{2^n\}$, whose n^{th} term is 2^n. The nth term of its difference sequence is therefore,

$$2^{n+1} - 2^n = (2 \times 2^n) - 2^n = 2^n,$$

so we have the result:

$$D(\{2^n\}) = \{2^n\}.$$

Thus, the difference sequence of $\{2^n\}$ is identical to its parent, and we have succeeded in understanding and explaining an empirically discovered property – a happy occurrence!

Likewise, for the sequence $\{3^n\}$, we have:

$$3^{n+1} - 3^n = (3 \times 3^n) - 3^n = (2 \times 3^n),$$

so:

$$D(\{3^n\}) = \{2 \cdot 3^n\}.$$

It is an easy matter now to deduce the following:

$$\begin{aligned} D_2(\{3^n\}) &= \{4 \cdot 3^n\}, \\ D_3(\{3^n\}) &= \{8 \cdot 3^n\}, \\ D_4(\{3^n\}) &= \{16 \cdot 3^n\}, \ \ldots . \end{aligned}$$

Thus, each difference sequence is exactly twice its 'parent', and our earlier observations now make sense – for instance, that the numbers get larger and larger as we go down the array, and that the initial numbers of the various levels are just the powers of 2.

The symbolic method works on $\{2^n + n\}$ as well. Here, the $(n+1)^{\text{th}}$ term is $2^{n+1} + n + 1$, so the nth term of its difference sequence is

$$(2^{n+1} + n + 1) - (2^n + n) = 2^n + 1,$$

and so

$$D(\{2^n + n\}) = \{2^n + 1\}.$$

Next, the $(n+1)^{\text{th}}$ term of $\{2^n + 1\}$ is $2^{n+1} + 1$, so the nth term of the difference sequence of this sequence is

$$(2^{n+1} + 1) - (2^n + 1) = 2^n,$$

and we finally have the result:

$$D_2(\{2^n + n\}) = \{2^n\}.$$

We have ultimately arrived at the observation made earlier!

Exercises

5.3.1 Consider the sequence $S = \{a^n\}$, where 'a' denotes some fixed number. Find formulas for the nth terms of the following sequences:
(i) $D(S)$ (ii) $D_2(S)$ (iii) $D_3(S)$ (iv) $D_4(S)$.

5.3.2 Compute the difference array for $\{(-1)^n + n^3\}$, then check whether results could have been anticipated via the symbolic method.

5.3.3 The numbers appearing in the difference array of $\{4^n\}$ have this in common: they are all divisors of the powers of some number. *Which number?*

5.3.4 Show that the numbers appearing in the array of differences of $\{5^n\}$ are precisely the numbers that are divisors of the powers of 10.

★ ★ ★

A famous sequence whose difference sequence is very much like the original sequence is the Fibonacci sequence; its 1st few terms are displayed below.

0,	1,	1,	2,	3,	5,	8,
13,	21,	34,	55,	89,	144,	233,
377,	610,	987,	1597,	2584,	4181,	6765.

The defining property of this sequence is that each term after the 1st two is the sum of the two preceding terms, e.g., $5 = 3 + 2$, $21 = 13 + 8$, and so on. Symbolically, if 'Fib' denotes the Fibonacci sequence, then

$$\begin{aligned} &\mathrm{Fib}_0 = 0, \quad \mathrm{Fib}_1 = 1, \quad \text{and} \\ &\mathrm{Fib}_n = \mathrm{Fib}_{n-1} + \mathrm{Fib}_{n-2} \quad \text{for all } n > 1. \end{aligned}$$

The seemingly simple definition hides the terrific rate at which the Fibonacci numbers grow. Here are a few numbers to illustrate the rate of growth of the sequence:

$$\begin{aligned} \mathrm{Fib}_{20} &= 6765, \\ \mathrm{Fib}_{40} &= 102334155, \\ \mathrm{Fib}_{60} &= 1548008755920, \\ \mathrm{Fib}_{80} &= 23416728348467685, \\ \mathrm{Fib}_{100} &= 354224848179261915075. \end{aligned}$$

It turns out that Fib_{200} is equal to

$$280571172992510140037611932413038677189525,$$

a 42-digit number!

⋆ ⋆ ⋆

We now describe a situation where the Fibonacci sequence arises quite naturally. We write the positive integers as sums of 1s and 2s in all the possible ways, order being considered relevant. Thus, 3 can be written in the following three ways:

$$1 + 1 + 1, \ 1 + 2, \ 2 + 1,$$

which, for convenience, we can write in the manner shown below:

111, 12, 21.

The corresponding list for 4 is

1111, 112, 121, 211, 22,

with five entries; and the list for 5 is

11111, 1112, 1121, 1211, 2111, 122, 212, 221,

with eight entries. The list for 6 has thirteen entries:

111111, 11112, 11121, 11211, 12111, 21111,
1122, 1212, 1221, 2112, 2121, 2211, 222,

while the list for 7 has twenty-one entries:

1111111, 111112, 111121, 111211, 112111,
121111, 211111, 11122, 11212, 11221, 12112,
12121, 12211, 21112, 21121, 21211, 22111,
1222, 2122, 2212, 2221.

This counting problem could arise as follows. I have before me a staircase with n steps, and I can climb only one or two steps at a time. In how many different ways can I reach the top of the staircase? A moment's reflection will reveal that the problem is essentially the same as the earlier one.

Let $f(n)$ denote the number of ways that the positive integer n can be expressed in this manner, as a sum of 1s and 2s. We list the values of $f(n)$ for $n = 1, 2, 3, \ldots$ in the table below.

n	1	2	3	4	5	6	7	8
$f(n)$	1	2	3	5	8	13	21	34

Note that, though we have lost the 1st two terms, $\{f(n)\}$ is essentially the Fibonacci sequence; in fact, $f(n) = \text{Fib}_{n+1}$ for each $n > 0$. You may want to verify that $f(8) = 34$ ($= \text{Fib}_9$) by actually listing the thirty-four possible ways of writing 8 as a sum of 1s and 2s.

⋆ ⋆ ⋆

The difference sequence of Fib is

$$\left\langle 1,\ 0,\ 1,\ 1,\ 2,\ 3,\ 5,\ 8,\ 13,\ \ldots \right\rangle,$$

which, if one ignores the initial '1', is the same as Fib. This suggests that Fib is an exponential sequence, for it bears the characteristic property of exponential sequences. If so, it is not at all obvious what the generating formula for the sequence might be.

It turns out that the guess is right: Fib *is* indeed an exponential sequence, but its generating formula contains non-integral numbers. More precisely, it is the sum of two exponential sequences, for each of which the generating formula contains non-integral numbers. The formulas, however, are such that the sums of the corresponding terms in the two sequences always turn out to be whole numbers. The exact formula is displayed below:

$$\text{Fib}_n = \frac{1}{\sqrt{5}}\left(\left(\frac{1+\sqrt{5}}{2}\right)^n - \left(\frac{1-\sqrt{5}}{2}\right)^n\right).$$

This may be rather more than you bargained for!

A nicer way to express the formula is the following. Let α and β be given by $\alpha = (\sqrt{5}+1)/2 \approx 1.618034$ and $\beta = -1/\alpha = -(\sqrt{5}-1)/2 \approx -0.618034$; then

$$\text{Fib}_n = \frac{\alpha^n - \beta^n}{\alpha - \beta}.$$

This is known as *Binet's formula*. Check that the values given by the formula for $n = 0, 1, 2, \ldots, 5$ agree with the values given earlier. Another formula is the following:

$$\text{Fib}_n = \text{the integer closest to } \frac{\alpha^n}{\sqrt{5}},$$

with α as defined above. This may come as a surprise. Let us check it for the values $n = 7$ and 8. We have $\sqrt{5} \approx 2.23607$, and

$$\frac{1.61803^7}{2.23607} \approx 12.9844, \quad \frac{1.61803^8}{2.23067} \approx 21.0091.$$

The integers closest to 12.9844 and 21.0091 are 13 and 21, respectively, and these are just the values of Fib_7 and Fib_8. There are other such formulas, and we shall find out more about them in Chapter 15.

⋆ ⋆ ⋆

A slight change in formula yields another important sequence, the *Lucas sequence*, $\{L_n\}$, which is defined by the rules $L_0 = 1$, $L_1 = 3$, $L_n = L_{n-1} + L_{n-2}$ for $n > 1$. It turns out that

$$L_n = \alpha^{n+1} + \beta^{n+1}.$$

The sequence starts as: 1, 3, 4, 7, 11, ...

⋆ ⋆ ⋆

We shall not say more here, except to add that the Fibonacci sequence is exceedingly rich in patterns, so much so that it even has a journal, *The Fibonacci Quarterly*, devoted entirely to it! You may enjoy discovering some of these patterns on your own. (For more on the Fibonacci sequence, you can refer to Chapter 15.)

Chapter 6

Conclusion

6.1 General comments

We have now seen how a study of the difference array of a sequence can shed light on its inner structure, particularly for sequences generated by polynomial and exponential expressions. Remarkably, the method can be used to find formulas for sums such as $1 + 2 + \cdots + n$, $1^2 + 2^2 + \cdots + n^2$ and $1^3 + 2^3 + \cdots n^3$, n being an arbitrary whole number.

It is important to realize, however, that many number sequences that arise in mathematics and science, quite naturally are too complex to analyze using such methods. Probably, the best example is the sequence of prime numbers:

$$P = \Big\langle 2,\ 3,\ 5,\ 7,\ 11,\ 13,\ 17,\ 19,\ 23,\ 29,\ 31,\ \ldots \Big\rangle.$$

Its sequence of 1st differences,

$$D(P) = \Big\langle 1,\ 2,\ 2,\ 4,\ 2,\ 4,\ 2,\ 4,\ 6,\ 2,\ \cdots \Big\rangle,$$

is, to all appearances, empty of patterns. The sequence of 2nd differences,

$$D_2(P) = \Big\langle 1,\ 0,\ 2,\ -2,\ 2,\ -2,\ 2,\ 2,\ -4,\ \ldots \Big\rangle$$

looks equally empty! Going down further does not seem to help. Is P, then, free of patterns; or are there patterns that are too deep to discern? It is not easy to answer this without further study.

We should add, though, that the prime number sequence will beat almost any technique used for the analysis of sequences. There is no one-variable polynomial function, say of the type $f(n) = n^2 + n + 41$ (which we came across earlier), that can generate the sequence of prime numbers. Nor is there any one-variable polynomial function that takes only prime number values. Here, we permit ourselves to skip some prime numbers, but the numbers taken should all be prime. (Of course, we exclude trivial examples in which the values taken by the function are all the same, e.g., $f(n) = 3$ for all n; this function certainly does take only prime number values, but in a trivial sense.) The fact that such polynomials do not exist, can be proved by algebraic methods. (See Chapter 11 for an outline of the proof.) We shall not discuss the matter further at this point, except to say that prime number theory *is* difficult, and it is not surprising that there are so many unresolved conjectures about primes. For instance, we have the twin-primes conjecture and the Goldbach conjecture, both of which are unresolved as of today despite decades of strenuous effort. We make a few observations about them below.

- The *twin-primes conjecture* concerns prime twins, i.e., primes that differ by 2, for instance: $(3, 5)$, $(5, 7)$, $(11, 13)$, $(101, 103)$, ... The available evidence suggests that there are infinitely many such twins, and it is conjectured that this is the case, but no proof has yet been found.
- *Goldbach's conjecture* states that every even integer greater than 2 can be written as the sum of two primes, e.g., $60 = 7 + 53$, $100 = 3 + 97$, $200 = 7 + 193$. It has been proved that every sufficiently large, odd integer can be written as the sum of *three* primes. This was demonstrated by the Russian mathematician, Vinogradov; it is a deep result, and it is very hard to prove. But it is still a long way from Goldbach's assertion.

⋆ ⋆ ⋆

More such examples can be given, and it is not difficult to construct sequences generated by some "nice" formula, yet for which the method of differences fails miserably. We give two such instances, with the respective difference arrays displayed alongside:

- $f(n) = \left[\sqrt{n^3}\right]$ (here $[x]$ denotes the largest integer not greater than x).

1		2		5		8		11		14		18		22
	1		3		3		3		3		4		4	
		2		0		0		0		1		0		
			−2		0		0		1		−1			

- $g(n) = \left[2^{n+\sqrt{n}} \right]$.

4		10		26		64		150		349
	6		16		38		86		199	
		10		22		48		113		
			12		26		65			
				14		39				

No pattern can be discerned in either array.

* * *

In general, the following is true. Given a sequence $\{f_n\}$, the method of differences is effective when f_n can be written as a *linear* expression in $f_{n-1}, f_{n-2}, \ldots$. For example, the Fibonacci sequence has $\text{Fib}_n = \text{Fib}_{n-1} + \text{Fib}_{n-2}$ (note that the expression on the right is linear in Fib_{n-1} and Fib_{n-2}), and in the exponential sequence $\{2^n\}$ the nth term is twice the $(n-1)^{\text{th}}$ term. Linearity means that no squared terms are present, nor cubes, square roots, cross-products of terms, and so on. Thus, the method of differences works in the instances listed below:

1. $f_n = 2f_{n-1}$;
2. $f_n = 2f_{n-1} - f_{n-2}$;
3. $f_n = f_{n-1} + f_{n-2} + f_{n-3}$;
4. $f_n = f_{n-1} + f_{n-4} + f_{n-7}$;
5. $f_n = f_{n-1} + f_{n-2} + f_{n-3} + \cdots + f_1$ (equivalently: $f_n = 2f_{n-1}$);
6. $f_n = f_{n-1} + 2f_{n-2} + 3f_{n-3} + \cdots + (n-1)f_1$ (this is equivalent to $f_n = 2f_{n-1} + f_{n-2} - f_{n-3}$);

but not in the following cases:

7. $f_n = (f_{n-1})^2$;
8. $f_n = f_{n-1}f_{n-2}$;
9. $f_n = (f_{n-1})^2 + (f_{n-2})^2$;
10. $f_n = f_1 f_{n-1} + f_2 f_{n-2} + f_3 f_{n-3} + \cdots$;
11. $f_n = (1/f_{n-1} + 1/f_{n-2})^{-1}$;

12. $f_n = \left[\sqrt{(f_{n-1})^2 + 1}\right]^2.$

The book by Markushevich (please refer to the reading list) has more on this topic, and is highly recommended.

Exercises

To conclude, here are a few problems where you are asked to count the number of objects of some kind using the techniques presented in this book – that is, on the basis of an observed pattern in some associated number sequence. Remember that when you do discover a pattern, you will have to decide for yourself whether or not the pattern is genuine, and whether it can be used to obtain a formula for the nth term.

★ ★ ★

6.1.1 *Lines On A Plane–I*

Mark n points on a plane, and draw all the possible lines joining these points. (Figure A.6.1 shows the case $n = 4$.) Assume that no three of the points are collinear.

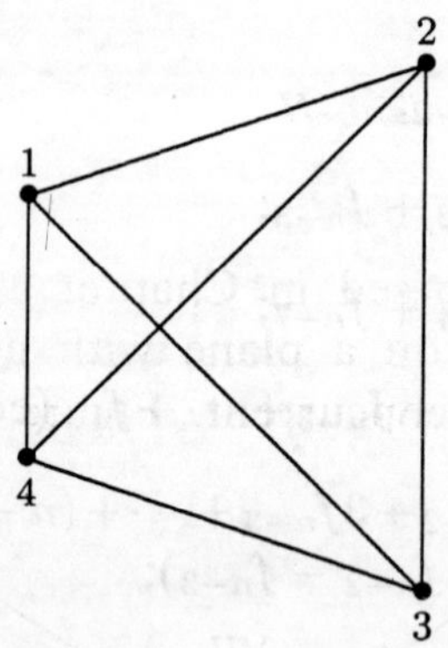

Figure A.6.1 *Lines on a plane–I*

Let $L(n)$ denote the total number of lines drawn. We have: $L(1) = 0$ (with one point no lines can be drawn), $L(2) = 1$ (with two points exactly one line can be drawn), $L(3) = 3$ (three points give rise to three lines), $L(4) = 6$ (as shown in the figure); and so on. Build up a table of values of $L(n)$, proceeding till $n = 8$ or so, then use the method of differences to work out a formula for $L(n)$.

6.1.2 *Lines On A Plane–II*

Draw n lines on a plane, such that no two lines are parallel to one another and no three are concurrent. (Figure A.6.2 shows the case $n = 4$.) Let $P(n)$ denote the number of points of intersection of these lines. We have:

$$P(1) = 0, \quad P(2) = 1, \quad P(3) = 3, \quad P(4) = 6,$$

and so on. Build up the sequence till $n = 8$ or so, then use the method of differences to work out a formula for $P(n)$.

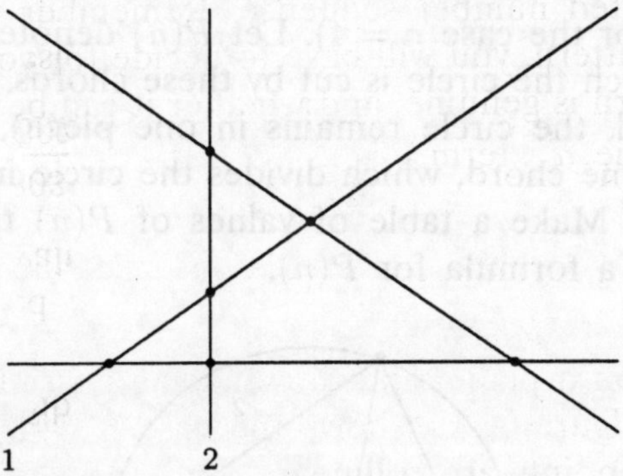

Figure A.6.2 *Lines on a plane–II*

6.1.3 *Lines On A Plane–III*

This problem was discussed in Chapter 1. As in the preceding problem, draw n lines on a plane with no two parallel to one another and no three concurrent. Figure A.6.3 shows the case $n = 3$.

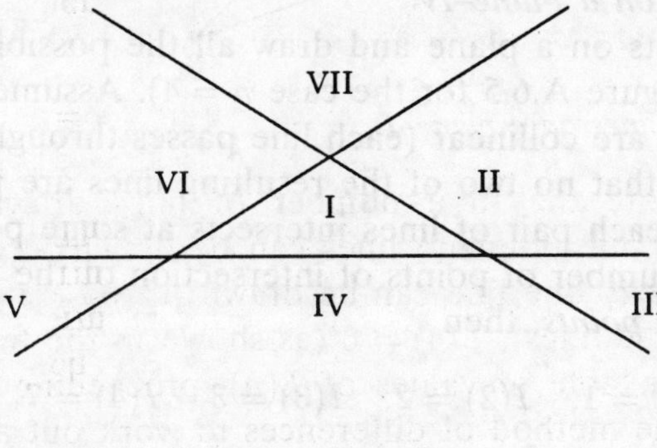

Figure A.6.3 *Regions created by drawing lines on a plane*

Let $R(n)$ denote the number of regions in the plane created by the lines. (Think of the plane as an infinite sheet of paper and the lines as cuts created by a knife.) We have:

$$R(0) = 1, \quad R(2) = 2, \quad R(3) = 4, \quad R(4) = 7,$$

and so on. Build up a table of values of $R(n)$ till $n = 8$ or so, then use the method of differences to work out a formula for $R(n)$.

6.1.4 *Chords in a Circle*

Mark n points on the periphery of a circle such that, when all the connecting chords are drawn, no three are concurrent (see Figure A.6.4 for the case $n = 4$). Let $P(n)$ denote the number of pieces into which the circle is cut by these chords. Thus, $P(1) = 1$ (with no chord, the circle remains in one piece), $P(2) = 2$ (two points create one chord, which divides the circle into two pieces), $P(3) = 4, \ldots$. Make a table of values of $P(n)$ till $n = 7$ or so, then work out a formula for $P(n)$.

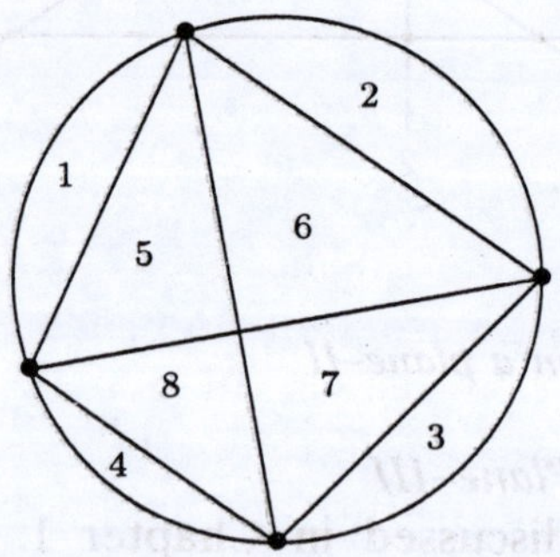

Figure A.6.4 *Chords in a circle*

6.1.5 *Lines on a Plane–IV*

Mark n points on a plane and draw all the possible lines joining them (see Figure A.6.5 for the case $n = 4$). Assume that no three of the points are collinear (each line passes through precisely two points), and that no two of the resulting lines are parallel to one another (so each pair of lines intersects at some point). Let $I(n)$ denote the number of points of intersection of the lines, *including the original n points*, then

$$I(1) = 1, \quad I(2) = 2, \quad I(3) = 3, \quad I(4) = 7, \quad \ldots.$$

(In Figure A.6.5., the 4 original points are marked with heavy dots, and the points obtained from the "next round", as points of intersection, are not so marked.)

Make a table of values of $I(n)$ for some more values of n (it will probably be too messy to proceed beyond $n = 8$), then work out a formula for $I(n)$.

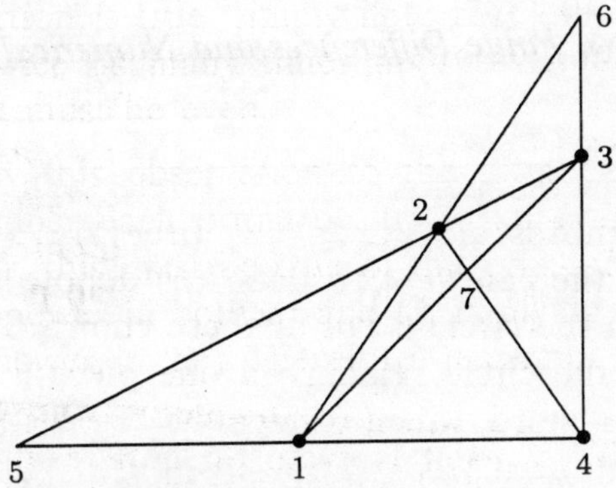

Figure A.6.5 *Lines on a plane–IV*

List of references

The references contained in this list are recommended for further study. Books [1] and [2] are quite advanced, but they contain an amazing wealth of material on all the aspects of algebra and came very highly recommended; they also have rich collections of problem. Books [3], [4] and [5] are about difference equations, and they describe how the generating formula of a sequence S can be deduced from information about the difference sequence $D(S)$ (perhaps in terms of a formula connecting S and $D(S)$); other aspects too are covered. Book [5] also describes how the method of differences can be used to find or estimate in-between values of a function from a partial list of values (this is called *interpolation*).

[1] Barnard & Child *Higher Algebra* (Macmillan).
This is an excellent source book on all aspects of old-fashioned algebra.

[2] Hall & Knight, *Higher Algebra* (S Chand).
This is another excellent source book that is very similar in content and style to the Barnard & Child book listed above.

[3] A Markushevich, *Recursion Sequences* (MIR Publishers, Moscow). This is part of the 'Little Mathematics Library'. It contains a rich fund of material on how difference equations arise and how they are solved.

[4] S Goldberg, *Difference Equations*.

[5] H C Saxena, *Finite Differences and Numerical Analysis*.

PART B

A Gallery of Sequences

The proof uses a curious idea first pioneered by Fermat, called "infinite descent". We suppose that there exists a prime number P of the form $1 + 4k$ which is *not* a sum of two squares. Now, it may be shown without much fuss that there must exist some multiple of P, say mP, which *is* a sum of two squares. According to our supposition, then, the least such m exceeds 1.

Now we produce, by ingenious but simple manipulations, a multiple of P, say $m'P$, which is smaller than mP but is itself a sum of two squares. This argument can now be applied to $m'P$ itself (because, by supposition, we have $m' > 1$), producing another such multiple, say $m''P$, and so on. Thus, we obtain an infinite sequence of positive multiples of P, each one smaller than the preceding one. But this is absurd; such a sequence cannot exist. Conclusion: the least m is 1, so P is indeed a sum of two squares. Lagrange applied the same idea to the four-squares problem and was able to get a complete proof of the statement. We shall not go into the details here, as they are quite involved.

Sums of squares — [illegible] we may proceed as follows. Let (a, b, c) [illegible]. Then $a^2 + b^2 = c^2$, and a, b, c share no common factor. It cannot happen that a and b are both odd; for if a and b were both odd, then this would lead to a^2 and b^2 both being of the form $1 + 4k$, leading, in turn, to c^2 being of the form $2 + 4k$. But no square is of this form; we have a contradiction! Therefore, exactly one of a and b is even, and we may suppose that it is a which is even. Now we write $b^2 = c^2 - a^2$, or

$$b^2 = (c - a)(c + a).$$

Since c and a have no factor in common, and c is odd whereas a is even, the numbers $c - a$, $c + a$ also share no common factor. Since their product is b^2, a square, it follows that $c - a$ and $c + a$ are both squares, say

$$c - a = (m - n)^2, \qquad c + a = (m + n)^2,$$

where m, n are integers with $m > n$. The two equations yield, by simple addition and subtraction respectively,

$$c = m^2 + n^2, \qquad a = 2mn.$$

Since $b^2 = (c - a)(c + a) = (m - n)^2(m + n)^2$, we get

$$b = (m - n)(m + n) = m^2 - n^2.$$

Part B

Introduction

As noted in the preface, this part of the book will be devoted to a tour of various sequences which are frequently encountered by mathematicians. So we have titled it "A Gallery of Sequences". The idea is to present various interesting features of each sequence: its particularities, idiosyncrasies, peculiarities, and so on.

Some readers may be disappointed to see the absence of detailed proofs. However this absence is to some extent inevitable, as the level of difficulty of many of these proofs lies far above the proposed level of this book. In such cases, we confine ourselves to making a few general remarks towards the end of the section, with proof techniques finding a brief mention. Of necessity these comments are terse and brief, but we hope that they will convey some idea of the underlying reasoning.

Chapter 7

The Sequence of Squares

7.1 Generating the squares

The squares may be generated from the sequence of odd numbers:

$$1, \quad 3, \quad 5, \quad 7, \quad 9, \quad 11, \quad \ldots$$

by cumulative addition. Thus, we have, $1 = 1^2$, $1+3 = 2^2$, $1+3+5 = 3^2$, $1+3+5+7 = 4^2$, and so on; in general, the sum of the 1st n odd numbers yields n^2. There are two ways of seeing why cumulative addition yields the squares in this manner.

The 1st way is through direct summation. Let S denote the sum of the 1st n odd numbers $(1, 3, 5, \ldots, 2n-3, 2n-1)$. We write S in two different ways by reversing the order in which the individual summands are listed. Thus, we have

$$\begin{aligned} S &= 1+3+5+7+\cdots+(2n-3)+(2n-1), \\ S &= (2n-1)+(2n-3)+\cdots+7+5+3+1. \end{aligned}$$

Adding the two rows column by column, we find that the sum of the two numbers in each column is the same:

$$1+(2n-1) = 2n, \quad 3+(2n-3) = 2n, \quad \ldots, \quad (2n-1)+1 = 2n.$$

Since each row has n terms, we see that $2S = n \times 2n$, so S is equal to n^2, as stated.

The other way is pictorial. Strictly speaking, this is only a *suggestion* of a proof, but it conveys the idea much more effectively that the algebraic proof presented above. In a way, it is a "proof without words".

We make a square grid of dots and group the dots into regions as shown below. The 1st region contains 1 dot, the 2nd region 3 dots, the 3rd contains 5 dots,. . . , and in general the nth region contains $2n-1$ dots. Since the total number of dots in the array is n^2, we see immediately that the sum of 1, 3, 5, . . . , $2n-1$ is n^2.

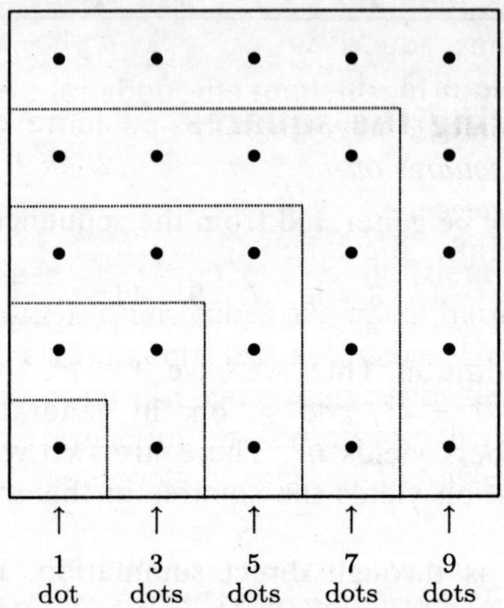

Figure B.7.1 *Square dot array*

* * *

Remark Devising proofs of this kind, where the use of words and symbols is kept to a minimum, or cut out altogether, requires great skill, and may be regarded as an art form.

7.2 Last digit

Consider the sequence formed by the units digits of the squares when written in normal decimal notation:

$$0, \quad 1, \quad 4, \quad 9, \quad 6, \quad 5, \quad 6, \quad 9, \quad 4, \quad 1, \quad 0, \quad \ldots$$

We observe that, between every two 0s, the same sequence of digits occurs: 1, 4, 9, 6, 5, 6, 5, 6, 9, 4, 1. The units digit thus exhibits a *periodic pattern*. Note that the digits 2, 3, 7 and 8 do not appear – they never occur as the units digits of a square.

More generally, we examine what remainders are possible when the squares are divided by different numbers. We find that under division by 3, remainders of 0 and 1 are obtained, but not a remainder of 2; so no square is of the form 2 plus a multiple of 3. For example, we have $\text{Rem}(4^2 \div 3) = 1$, $\text{Rem}(5^2 \div 3) = 1$, $\text{Rem}(6^2 \div 3) = 0$, and so on.

Under division by 4, too, the only remainders possible are 0 and 1; for example, $\text{Rem}(6^2 \div 4) = 0$, $\text{Rem}(7^2 \div 4) = 1$, $\text{Rem}(8^2 \div 4) = 0$, and so on. Thus, squares of the type 2 plus a multiple of 4, or 3 plus a multiple of 4, do not exist.

These statement may be expressed more concisely as follows: *There are no squares of the forms* $3k+2$, $4k+2$ *or* $4k+3$, *where* k *is a positive integer.*

More generally, the following is true:

> *For each divisor* d, *there exist some numbers* i *within the set* $\{1, 2, \ldots, d-2, d-1\}$, *such that there are no squares of the form* i *plus a multiple of* d.

When $d = 4$ the values of i are 2 and 3, as mentioned above; when $d = 5$ the values of i are again 2 and 3 (implying that there are no squares of the forms $5k+2$ and $5k+3$, thus, every square is of one of the forms $5k$, $5k+1$, $5k+4$); when $d = 8$ the values of i are 2, 3, 5, 6 and 7 (thus, every square is of one of the forms $8k$, $8k+1$, $8k+4$); and so on.

7.3 Coin changing using squares

Imagine that currency is available only in the amounts 1, 4, 9, 16, ... Thus, we have a Re 1/- coin, a Rs 4/- coin, a Rs 9/- note, a Rs 16/- note, a Rs 25/- note, and so on (we do not distinguish between coins and notes). This means infinitely many different currency denominations—something that the Reserve Bank would not be too happy about! The question we ask here is: *can every amount be paid out using this currency system?* Of course this is possible—trivially so, since we have a unit available (that is, the Re 1/- coin). But something much more interesting can be said: *we will never need more than four notes!* No matter what amount has to be paid out, we can always do so using four notes or less. For example, Rs 60/- can be paid out as $60 = 49 + 9 + 1 + 1$; Rs 73/- can be paid out as $73 = 36 + 36 + 1$, and so on.

It is not easy to show why this must be so. The fact that every positive integer is a sum of four or fewer squares was probably known to Fermat and it was certainly known to Euler; but he was unable to show *why* the statement is true. Eventually, it was Lagrange who proved the result.

If we were told that we could never use more than *two* notes in any single transaction, then a great many transactions would become impossible. For instance, we would never be able to pay out amounts such as Rs 3/-, or 7/-, or 11/-, or 15/-, ..., because the numbers 3, 7, 11, 15, ..., are neither squares nor sums of two squares. Let us see why this is so.

Observe that these numbers are all of the form $3+4k$, where k is a positive integer. We have already seen that no square is of this form. We now show that a sum of two squares cannot be of this form either. All squares are of the form $4k$ or $1+4k$, so writing u and v respectively for the "k-values" of the two squares, a sum of two squares must necessarily be of one of the following forms:

$$4u+4v, \quad \text{or} \quad 4u+(1+4v), \quad \text{or} \quad (1+4u)+(1+4v).$$

Writing m for $u+v$, the numbers take the forms $4m$, $1+4m$ and $2+4m$, respectively. Note that the form $3+4m$ is absent from the list. This means that numbers of the form $3+4m$ cannot occur as sums of two squares.

However, we can say more. There are also numbers which are of the form $1+4m$ which are not sums of two squares – for instance, 21 and 33 (as can be checked by hand, the numbers involved being small). There is a nice explanation behind this, and it can be seen by factorizing the numbers. We find that $21 = 3 \times 7$ and $33 = 3 \times 11$; observe that each prime factor is of the "bad" type. We may guess from this that the number $7 \times 11 = 77$ too is not a sum of two squares, and we find that this is true. Indeed, the following can be said:

> *If the positive integer N has prime factors of the type 3 plus a multiple of 4, and if it has an* <u>*odd*</u> *number of any one such factor, then it is not a sum of two squares; and it is only such numbers which are not sums of two squares.*

For example, consider the number 63, whose prime factorization is $3^2 \times 7$. The factors 3 and 7 are both of the type "3 plus a multiple of 4". Of these, 3 is repeated twice, while 7 occurs singly.

The single occurrence of 7 is enough to classify 63 as a number which is not a sum of two squares.

Likewise, consider the number $4455 = 3^4 \times 5 \times 11$. The prime factors 3 and 11 are of the type "3 plus a multiple of 4" and, of these, 3 occurs to an even power whereas 11 occurs to an odd power. So, 4455 is not a sum of two squares. On the other hand, the number $405 = 3^4 \times 5$ is a sum of two squares; indeed, $405 = 324 + 81 = 18^2 + 9^2$.

So the numbers which are not sums of two squares are: 3, 6, 7, 11, 12, 14, 15, 19, 21, 22, 24, 27, 28, 30, 31, 33, 35, ... The reader should check that the property described above holds true for each of these numbers.

7.4 Sums of squares

Can the sum of two squares be yet another square? This is certainly possible, for instance,

$$3^2 + 4^2 = 9 + 16 = 25 = 5^2,$$
$$5^2 + 12^2 = 25 + 144 = 169 = 13^2,$$
$$7^2 + 24^2 = 49 + 576 = 625 = 25^2,$$

and so on. Here is a simple way to generate infinitely many such instances. Take any odd (positive) integer n and find the sum of the fractions $1/n$ and $1/(n+2)$. Let the sum be a/b, where a, b are positive integers; then $a^2 + b^2$ is a square. For example,

(a) Let $n = 5$; then we have

$$\frac{1}{5} + \frac{1}{7} = \frac{12}{35},$$

and we have $12^2 + 35^2 = 144 + 1225 = 1369 = 37^2$.

(b) Or let $n = 9$; then

$$\frac{1}{9} + \frac{1}{11} = \frac{20}{99},$$

and we find that $20^2 + 99^2 = 400 + 9801 = 10201 = 101^2$.

However, we must point out here that this method does not exhaust all the possibilities; there are instances of two squares adding up to another square which are not captured by this formula. For instance, we have

$$33^2 + 56^2 = 1089 + 3136 = 4225 = 65^2,$$

and this is not obtainable by our formula.

Whole number triples (a, b, c) which have the property that $a^2 + b^2 = c^2$ are called *Pythagorean triples*, in honour of the mathematician-philosopher Pythagoras, because of the obvious connection with the theorem of Pythagoras: *If* $\triangle ABC$ *is right-angled at C, and its side lengths are* a, b, c, *then* $c^2 = a^2 + b^2$. The phrase "Pythagorean triple" is often shortened to PT, and if the terms in the PT have no factors in common, then the triple is called a "primitive PT," or PPT for short. So $(3, 4, 5)$ and $(5, 12, 13)$ are examples of PPTs, but not $(6, 8, 10)$.

⋆ ⋆ ⋆

Are there instances of *three* squares whose sum is a square? Certainly this is possible; for example,

$$2^2 + 3^2 + 6^2 = 4 + 9 + 36 = 49 = 7^2.$$

Here, too, infinitely many such instances can be generated by a formula: let a be any positive integer, $b = a + 1$ and $c = ab$; then $a^2 + b^2 + c^2$ is a square. For example,

(a) Let $a = 1$; then $b = 2$ and $c = 2$, and we have,

$$1^2 + 2^2 + 2^2 = 1 + 4 + 4 = 9 = 3^2.$$

(b) Let $a = 5$; then $b = 6$ and $c = 30$, and we have,

$$5^2 + 6^2 + 30^2 = 25 + 36 + 900 = 961 = 31^2.$$

(c) Let $a = 10$; then $b = 11$ and $c = 110$, and we have,

$$10^2 + 11^2 + 110^2 = 12321 = 111^2.$$

As earlier, however, this formula does not exhaust all the possibilities. For instance, we have $1^2 + 4^2 + 8^2 = 81 = 9^2$, but this is not obtainable using our formula.

⋆ ⋆ ⋆

Can the sum

$$1^2 + 2^2 + 3^2 + \cdots + n^2$$

of the 1st n squares be a square? Certainly, if we put $n = 1$. Is there a more interesting instance? Yes, if we put $n = 24$. We can check this as follows. We know from Part A that the sum of the numbers $1^2, 2^2, \ldots, (n-1)^2, n^2$ is

$$\frac{n(n+1)(2n+1)}{6}.$$

When $n = 24$, the formula yields $(24 \times 25 \times 49)/6$ which equals 4900 or 70^2, a square.

So the sum of the first n squares is itself a square when $n = 1$ and 24. When else does it happen? The answer is: *never!* This can be shown mathematically, but unfortunately the proof is very difficult.

7.5 Squares in arithmetic progression

Can we find three squares in arithmetic progression (AP for short)?

By an AP, we mean a sequence which changes by a fixed amount at each step; the amount is called the *common difference* of the AP. Two examples are given below.

(a) The odd numbers (1, 3, 5, 7, . . .) form an AP with a common difference 2, as do the even numbers (2, 4, 6, 8, . . .).

(b) The sequence of numbers of the form $1 + 4k$, namely $\langle 1, 5, 9, 13, 17, \ldots \rangle$, is an AP with a common difference 4.

Can we find a three-term AP composed only of squares? Indeed we can, and there are infinitely many instances:

$$\begin{array}{llll} 7^2, & 13^2, & 17^2 & (\text{that is, } 49,\ 169,\ 289); \\ 7^2, & 17^2, & 23^2 & (\text{that is, } 49,\ 289,\ 529); \\ 17^2, & 25^2, & 31^2 & (\text{that is, } 289,\ 625,\ 961); \end{array}$$

and so on. These instances are provided by the triples (X, Y, Z), given by the following formula:

$$X = n^2 - 2n - 1, \quad Y = n^2 + 1, \quad Z = n^2 + 2n - 1.$$

To avoid having a common factor running through all members of the triple, we should choose an even number for n. Examples:

(a) With $n = 3$ we get the triple $(2, 10, 14)$, which has the factor 2 common to all the members. Dividing through by 2, we get the triple $(1, 5, 7)$. The squares of these numbers (namely, 1, 25, 49) do indeed form an AP.

(b) With $n = 10$ we get the triple $(79, 101, 119)$. The squares of these numbers are 6241, 10201, 14161 respectively, and they do indeed form an AP, for $10201 - 6241 = 3960$ and $14161 - 10201 = 3960$.

[Note that for (b), a nicer way to do the check is by using the formula $a^2 - b^2 = (a - b)(a + b)$. We have

$$\begin{aligned} 101^2 - 79^2 &= (101 - 79) \times (101 + 79) = 22 \times 180, \\ 119^2 - 101^2 &= (119 - 101) \times (119 + 101) = 18 \times 220, \end{aligned}$$

and it is obvious that $22 \times 180 = 18 \times 220$.]

⋆ ⋆ ⋆

There is a nice connection between instances of three squares in AP and Pythagorean triples. Thus, if (a, b, c) is a PT (that is, $a^2 + b^2 = c^2$) with $a < b$, then the triple (X, Y, Z) where

$$X = b - a, \quad Y = c, \quad Z = b + a,$$

gives an instance of "three squares in AP" – the numbers X^2, Y^2, Z^2 are in AP.

Examples

(a) the PPT $(3, 4, 5)$ yields the triple $(1, 5, 7)$, and 1^2, 5^2 and 7^2 are in AP;

(b) the PPT $(7, 24, 25)$ yields the triple $(17, 25, 31)$, and 17^2, 25^2 and 31^2 are in AP.

Both these instances have been encountered earlier in the chapter.

The claim just made is easy to verify. Let (a, b, c) be a PT; then $a^2 + b^2 = c^2$. We need to check whether $Y^2 - X^2 = Z^2 - Y^2$; equivalently, whether $X^2 + Z^2 = 2Y^2$. But this is easy:

$$\begin{aligned} X^2 &= (b - a)^2 = b^2 - 2ab + a^2, \\ Z^2 &= (b + a)^2 = b^2 + 2ab + a^2, \\ Y^2 &= c^2, \end{aligned}$$

so $X^2 + Z^2 = 2(a^2 + b^2) = 2c^2 = 2Y^2$, as desired. The steps may be reversed, showing that from a three-squares triple we obtain a PT. That is, if X^2, Y^2, Z^2 are in AP with $X < Y < Z$, then the triple (a, b, c) where

$$a = \frac{Z - X}{2}, \quad b = \frac{Z + X}{2}, \quad c = Y,$$

is a PT. If the triple (X, Y, Z) is primitive, with no common factors dividing all three numbers, then (a, b, c) is a PPT.

⋆⋆⋆

Having considered the case of three squares in AP, it is natural to ask whether there are instances of *four* distinct squares in AP. Surprise! – it turns out that there are no such instances to be found! However, this claim is very hard to prove, and we shall not attempt do so here.

⋆⋆⋆

We saw earlier that the AP

$$1,\ 3,\ 5,\ 7,\ 9,\ 11,\ 13,\ 15,\ \ldots$$

has the feature that all its partial sums are squares. Are there other APs whose partial sums are all squares? Remarkably, the above AP is one of a kind. Specifically, the following may be said: *if an* AP *is such that all its partial sums are squares, then it may be obtained by multiplying the* AP *1, 3, 5, 7, 9, 11, ... by some square number.* For example, the AP obtained by multiplying the above AP by 4, namely

$$4,\ 12,\ 20,\ 28,\ 36,\ 44,\ 52,\ 60,\ \ldots,$$

certainly has the feature we have in mind, and so does the one obtained if we multiply instead by 9:

$$9,\ 27,\ 45,\ 63,\ 81,\ 99,\ 117,\ 135,\ \ldots;$$

and so on. It can be shown mathematically that *every* such AP is of the form

$$k^2,\ 3k^2,\ 5k^2,\ 7k^2,\ 9k^2,\ 11k^2,\ \ldots,$$

for some integer k.

7.6 One square twice another?

Can one square be twice another square (other than the trivial instance $0^2 = 2 \times 0^2$)? We shall show by a neat argument that this cannot happen. The argument is a very ancient one – it is from Euclid's text, *The Elements*. Here is how it goes.

We start by supposing that it *is* possible to find such pairs of squares, and that a^2, b^2 are such a pair; so a, b are positive integers, and $a^2 = 2b^2$. We may suppose, without losing anything, that a and b have no factors in common (if there were any common factors, they may simply be "cancelled" away). We now reason as follows. Since $a^2 = 2b^2$, it follows that a^2 is an even number and therefore, that a itself is even; say $a = 2c$, where c is some positive integer. From this it follows that $a^2 = 4c^2$ and therefore, that $4c^2 = 2b^2$, or $b^2 = 2c^2$. The last equality implies that b^2 is an even number, and this means that b itself is even. So both a and b are even numbers; that is, they share the factor 2.

We have found a self-contradiction! – we had assumed at the start that a and b have no factors in common, and now we find that they share the factor 2. This self-contradiction shows that the equation $a^2 = 2b^2$ cannot hold good in the case of positive integers.

Mathematicians express the finding we have just made by stating that *the square root of 2 is irrational*. A "rational" number is of the form a/b where a and b are integers (with b non-zero), so the irrationality of a given number z means that the equation $z = a/b$ is not possible to solve in integers a and b. If z is the square root of 2, then the equation reads:

$$\sqrt{2} = \frac{a}{b},$$

or, after squaring and clearing fractions, $a^2 = 2b^2$. This is the same as the equation considered above, so the two statements are considered to be the same.

★ ★ ★

There are other possible approaches to show that $\sqrt{2}$ is irrational; here are two short proofs.

- Let a and b be positive integers. We shall show that the equation $a^2 = 2b^2$ cannot hold good.

The proof rests on the following, very easy result. *If p is a prime number and n is a positive integer, then p occurs to an even power in the prime factorization of n^2.* (Note that the statement is valid even if p does not divide n^2, for the exponent is 0 in this case and 0 is even!) To see why the assertion is true, simply note that every prime number occurs twice as many times in n^2 as it does in n, so its exponent must be even.

We apply this observation to the numbers a^2 and b^2. In each number each prime occurs to an even power and, in particular, this must be true of the prime 2. So 2 occurs to an even power in both a^2 and b^2, implying that it occurs to an *odd* power in $2b^2$. Therefore, it cannot possibly happen that $a^2 = 2b^2$.

- The second proof rests on the following principle: *Any set of positive integers has a smallest element.*

  Suppose that $\sqrt{2}$ is a rational number. Then there exist positive integers n such that $n\sqrt{2}$ is an integer. Let a be the *least* such integer. Then $a > 0$, $a\sqrt{2}$ is an integer; and, if $0 < b < a$, then $b\sqrt{2}$ is *not* an integer.

  Let $b = a\sqrt{2} - a$. Then $b > 0$ (because $\sqrt{2} > 1$), and b is an integer (it is one integer minus another). Also, it is less than a, because $\sqrt{2} - 1$ is less than 1. So b is a positive integer and it is smaller than a. Next,

  $$b\sqrt{2} = (a\sqrt{2} - a) \times \sqrt{2} = 2a - a\sqrt{2},$$

  so $b\sqrt{2}$ is an integer. We have found a positive integer b, which is smaller than a and for which $b\sqrt{2}$ is an integer! This contradicts the choice made right at the start. We conclude, therefore, that $\sqrt{2}$ is irrational.

As it happens, the square roots of 3 and 5 are irrational too. So it is not possible to find two squares such that one is 3 times the other, or 5 times the other, except trivially. The proofs are similar to the one given above.

★ ★ ★

Thus, it is impossible for one square to be twice another square, except trivially. This being so, we need not waste any more-time looking for such instances. Instead, we could look for "close misses" – instances where one square is *nearly* twice another square. The closest miss possible is ± 1, so what we are asking for is solutions in positive integers to the equations

$$a^2 - 2b^2 = 1, \quad a^2 - 2b^2 = -1.$$

These equations turn out to be rich subjects for investigation. Each has an infinite family of solutions! For the 1st equation, the family is shown in the following table (the dots show that the sequences continue):

a	1	3	17	99	577	3363	...
b	0	2	12	70	408	2378	...

Thus, we have, $3^2 - (2 \times 2^2) = 9 - 8 = 1$, $17^2 - (2 \times 12^2) = 289 - 288 = 1$, and so on.

It is not too hard to find a pattern in the two rows. If the nth terms in the two rows (above) are denoted by a_n and b_n respectively, then we find that

$$\left.\begin{array}{lll} a_n & = & 6a_{n-1} - a_{n-2} \\ b_n & = & 6b_{n-1} - b_{n-2} \end{array}\right\}.$$

In the language of Chapters 3 and 4 (Part A), both the a- and b-sequences are exponential in nature.

For the equation $a^2 - 2b^2 = -1$, we find the following family of solutions.

a	1	7	41	239	1393	8119	...
b	1	5	29	169	985	5741	...

Thus, we have, $7^2 - (2 \times 5^2) = 49 - 50 = -1$, $41^2 - (2 \times 29^2) = 1681 - 1682 = -1$, and so on. The same recursive law applies here: $a_n = 6a_{n-1} - a_{n-2}$, $b_n = 6b_{n-1} - b_{n-2}$.

⋆ ⋆ ⋆

We may similarly investigate the "near-miss" version of the other problem – that of finding one square which is three times another. The problem now is to solve the equations $a^2 - 3b^2 = \pm 1$ in positive integers. With the minus sign we get no solution (because no square is of the form $3k - 1$); but with the plus sign we get infinitely many solutions. These are displayed in the following table, with the dots indicating, as earlier, that the sequences continue.

a	1	2	7	26	97	362	...
b	0	1	4	15	56	209	...

Thus, we have, $2^2-(3\times 1^2)=4-3=1$, $26^2-(3\times 15^2)=576-575=1$, and so on.

As earlier, the two rows have a simple pattern to them. If the nth terms in the two rows are denoted by a_n and b_n, respectively, then we find that

$$\left.\begin{array}{rcl} a_n & = & 4a_{n-1}-a_{n-2} \\ b_n & = & 4b_{n-1}-b_{n-2} \end{array}\right\}.$$

7.7 Postscript on proofs

A few comments will be made here concerning the proof techniques used in making the claims described above.

Last digits of squares The mathematics here is straightforward. If the divisor is 5, we observe that any integer n is of one of the following forms,

$$5k, \quad 5k\pm 1, \quad 5k\pm 2,$$

where k is an integer. Squaring, we see that n^2 is of one of the following forms:

$$25k^2, \quad 25k^2\pm 10k+1, \quad 25k^2\pm 10k+4.$$

Writing m for $5k^2$ or $5k^2\pm 2k$ as the case may be, we see that all squares are of one of the following three forms:

$$5m, \quad 5m+1, \quad 5m+4.$$

This agrees with the claim made earlier. The other claims are demonstrated in the same way.

Coin changing Lagrange's proof of the four-squares theorem, that every positive integer is a sum of four squares, is very similar to the proof given by Euler himself to another such theorem—the two-squares theorem of Fermat, which states that *any prime number of the form* $1+4k$ *is a sum of two squares.*

The proof uses a curious idea first pioneered by Fermat, called "infinite descent". We suppose that there exists a prime number P of the form $1+4k$ which is *not* a sum of two squares. Now, it may be shown without much fuss that there must exist some multiple of P, say mP, which *is* a sum of two squares. According to our supposition, then, the least such m exceeds 1.

Now we produce, by ingenious but simple manipulations, a multiple of P, say $m'P$, which is smaller than mP but is itself a sum of two squares. This argument can now be applied to $m'P$ itself (because, by supposition, we have $m' > 1$), producing another such multiple, say $m''P$, and so on. Thus, we obtain an infinite sequence of positive multiples of P, each one smaller than the preceding one. But this is absurd; such a sequence cannot exist. Conclusion: the least m is 1, so P is indeed a sum of two squares. Lagrange applied the same idea to the four-squares problem and was able to get a complete proof of the statement. We shall not go into the details here, as they are quite involved.

Sums of squares To generate PPTs we may proceed as follows. Let (a, b, c) be a PPT. Then $a^2 + b^2 = c^2$, and a, b, c share no common factor. It cannot happen that a and b are both odd; for if a and b were both odd, then this would lead to a^2 and b^2 both being of the form $1+4k$, leading, in turn, to c^2 being of the form $2+4k$. But no square is of this form; we have a contradiction! Therefore, exactly one of a and b is even, and we may suppose that it is a which is even. Now, we write $b^2 = c^2 - a^2$, or

$$b^2 = (c-a)(c+a).$$

Since c and a have no factor in common, and c is odd whereas a is even, the numbers $c-a, c+a$ also share no common factor. Since their product is b^2, a square, it follows that $c-a$ and $c+a$ are both squares, say

$$c-a = (m-n)^2, \qquad c+a = (m+n)^2,$$

where m, n are integers with $m > n$. The two equations yield, by simple addition and subtraction respectively,

$$c = m^2 + n^2, \qquad a = 2mn.$$

Since $b^2 = (c-a)(c+a) = (m-n)^2(m+n)^2$, we get

$$b = (m-n)(m+n) = m^2 - n^2.$$

So the scheme

$$\begin{aligned} a &= 2mn, \\ b &= m^2 - n^2, \\ c &= m^2 + n^2 \end{aligned}$$

generates a PT. If we choose m and n to be of opposite parity (that is, one even and the other odd), then (a, b, c) turns out to be a PPT, and this scheme generates *all* the PPTs.

Three squares in arithmetic progression The proof technique used here is very similar to that used in finding PPTs. Let a, b, c be positive integers, such that a^2, b^2, c^2 are in AP. Then

$$b^2 - a^2 = c^2 - b^2, \quad \therefore\ a^2 + c^2 = 2b^2.$$

The second relation shows that a and c have the same parity, that is, they are either both odd or both even. Since all squares are of the type $4k$ or $1 + 4k$, we find that b too has the same parity as a and c. Thus, a, b, c are either all odd or all even. If the second possibility holds, then we can factor out 2 from all of them and obtain another instance of three squares in AP.

We shall not derive the general solution to the problem here, as we have already found the general formula for generating PTs and pointed out the close connection between these two problems; we only quote the general formula here. Choose two positive integers m and n of opposite parity (one odd, the other even) and with no factors in common, and let

$$\begin{aligned} a &= m^2 - 2mn - n^2, \\ b &= m^2 + n^2, \\ c &= m^2 + 2mn - n^2. \end{aligned}$$

Then (a^2, b^2, c^2) are in AP, and this scheme generates *all* instances of three squares in AP.

For example, the choice $m = 5$, $n = 2$ yields $a = 1$, $b = 29$, $c = 41$, and it is easily checked that $1^2, 29^2, 41^2$ are in AP.

One square nearly twice another We shall now make a few brief remarks in brief on the solutions in integers of the equation $a^2 - 2b^2 = 1$. Specifically, we shall show that every pair in the array

a	1	3	17	99	577	3363	...
b	0	2	12	70	408	2378	...

(as generated by the recursion given earlier) provides a solution to the equation. We shall use a proof technique known as *proof by mathematical induction*. The proof is quite intricate, as will be seen below.

We start with the pairs $(a_1, b_1) = (1, 0)$ and $(a_2, b_2) = (3, 2)$, and generate more pairs using the recursive law:

$$a_n = 6a_{n-1} - a_{n-2}, \quad b_n = 6b_{n-2} - b_{n-2} \quad (n > 2).$$

Our objective will be to show that the pairs (a_n, b_n) all fit the given equation. Certainly this is so for $n = 1$ and $n = 2$.

First, observe that

$$(1 \times 3) - (2 \times 0 \times 2) = 3, \quad (3 \times 17) - (2 \times 2 \times 12) = 3, \quad \ldots.$$

Claim The pattern continues; that is,

$$(a_{n-1} \times a_n) - (2 \times b_{n-1} \times b_n) = 3 \quad \text{for all } n > 1.$$

To show this, we inductively *assume* that the two relations

(a) $a^2 - 2b^2 = 1$,

(b) $aa' - 2bb' = 3$,

hold for all pairs (a, b) and for all adjacent pairs (a, b) and (a', b') respectively, till the nth pair. We now have

$$\begin{aligned} a_n a_{n+1} - 2b_n b_{n+1} &= a_n(6a_n - a_{n-1}) - 2b_n(6b_n - b_{n-1}) \\ &= 6(a_n^2 - 2b_n^2) - (a_{n-1}a_n - 2b_{n-1}b_n) \\ &= (6 \times 1) - 3 = 3. \end{aligned}$$

So $aa' - 2bb' = 3$ for the *next* adjacent pair too. Next, we check whether $a^2 - 2b^2 = 1$ for the $(n+1)^{\text{st}}$ pair. Recall that $a_{n+1} = 6a_n - a_{n-1}$ and $b_{n+1} = 6b_n - b_{n-1}$. We have therefore,

$$\begin{aligned} a_{n+1}^2 - 2b_{n+1}^2 &= (6a_n - a_{n-1})^2 - 2\,(6b_n - b_{n-1})^2 \\ &= 36(a_n^2 - 2b_n^2) + (a_{n+1}^2 - 2b_{n+1}^2) \\ &\quad -12(a_n a_{n+1} - 2b_n b_{n+1}) \\ &= 36 + 1 - (12 \times 3) = 1. \end{aligned}$$

So $a^2 - 2b^2 = 1$ for the $(n+1)^{\text{st}}$ pair too. Since this logic can be continued indefinitely, it follows that the two relations $aa' - 2bb' = 3$ and $a^2 - 2b^2 = 1$ continue to hold good for the entire sequence.

(It is interesting to see how closely the two relations work with one another – it is rather like a marriage between the two!)

Using more reasoning of this type, we can show that *every* solution of the equation $a^2 - 2b^2 = 1$ belongs to the sequence described above. However, we shall not go into that question here.

Four squares in arithmetic progression The task of showing that we cannot find four distinct squares in AP is a most formidable one, and we shall not enter into a discussion of the details here. The interested reader is referred to the book by L J Mordell, *Diophantine Equations*.

Exercises

7.7.1 Show that there are no squares of the forms $5k+2$, $5k+3$.

7.7.2 Show that there are no squares of the form $7k-1$.

7.7.3 Show that all odd squares leave a remainder of 1 when divided by 8.

7.7.4 Find the possible remainders left when the odd squares are divided by 16.

7.7.5 Fermat's Two Square Theorem states the following: *a prime of the form* $4k+1$ *can be written as a sum of two squares, and in precisely one way.* Verify this statement for the following five primes:

$$73, \quad 109, \quad 337, \quad 449, \quad 1009.$$

7.7.6 Investigate the following problem: *Which integers can be written as differences of two squares?* Example: The integers 3, 8 and 21 are of this kind, because

$$3 = 2^2 - 1^2, \quad 8 = 3^2 - 1^2, \quad 21 = 5^2 - 2^2.$$

But not every integer has this property; for instance, 6 is not a difference of two squares, nor is 14.

Chapter 8

The Sequence of Cubes

We have said quite a lot about the sequence of squares! In contrast, we shall say much less about the sequence of cubes:

$$\langle 1,\ 8,\ 27,\ 64,\ 125,\ 216,\ 343,\ \ldots \rangle.$$

This is not because the sequence is deficient in features of interest – far from it! Rather, cubes are much harder to work with than squares, so the results are less accessible. However, here too we shall be able to point out numerous curiosities.

There will be no postscript on "Proofs" in this chapter, for the simple reason that the proofs of most of the assertions are much too difficult.

8.1 Cumulative sums

The first property we focus on deals with the cumulative sums 1^3, 1^3+2^3, $1^3+2^3+3^3$, ... This is what we find after doing some computations:

$$\begin{aligned} 1^3 &= 1 \\ 1^3+2^3 &= 9 \\ 1^3+2^3+3^3 &= 36 \\ 1^3+2^3+3^3+4^3 &= 100 \\ 1^3+2^3+3^3+4^3+5^3 &= 225. \end{aligned}$$

We quickly notice that the numbers at the right are all squares. Moreover, we find the following:

$$\begin{aligned} 1 &= 1^2 \\ 9 &= (1+2)^2 \\ 36 &= (1+2+3)^2 \end{aligned}$$

$$\begin{aligned} 100 &= (1+2+3+4)^2 \\ 225 &= (1+2+3+4+5)^2. \end{aligned}$$

The pattern is unmistakable, and we conjecture that

$$1^3+2^3+3^3+\cdots+n^3=(1+2+3+\cdots+n)^2.$$

This conjecture turns out to be true. We may prove the statement using mathematical induction. We already know that

$$1+2+3+\cdots+n=\frac{n(n+1)}{2}.$$

Therefore, what we need to prove is the identity

$$1^3+2^3+3^3+\cdots+n^3=\frac{n^2(n+1)^2}{4}.$$

We know that this statement is true for $n = 1$, 2, 3, 4 and 5. Let us assume that is true for some value $n = k$. Then,

$$1^3+2^3+3^3+\cdots+k^3=\frac{k^2(k+1)^2}{4}.$$

Add $(k+1)^3$ to both sides. The right side becomes

$$\begin{aligned} \frac{k^2(k+1)^2}{4}+(k+1)^3 &= \frac{k^2(k+1)^2+4(k+1)^3}{4} \\ &= \frac{k^4+6k^3+13k^2+12k+4}{4} \\ &= \frac{(k^2+3k+2)^2}{4} \\ &= \left(\frac{(k+1)(k+2)}{2}\right)^2 \\ &= [1+2+3+\cdots+k+(k+1)]^2. \end{aligned}$$

So, *if* the relation

$$1^3+2^3+3^3+\cdots+n^3=\frac{n^2(n+1)^2}{4}$$

holds for some value $n = k$, then it also holds for $n = k + 1$ and, therefore, also for $n = k + 2$ and, therefore, also for $n = k + 3$, and so on. We already know that the statement is true for $n = 1$ and $n = 2$, so we see that it holds for every positive integer n.

⋆ ⋆ ⋆

The property just proved may be expressed in words as follows: *The list of numbers 1, 2, 3, ..., n has the property that the sum of the cubes of these numbers equals the square of their sum.* The reader may wonder whether there are other lists that have this property. Indeed there are, and infinitely many instances may be generated according to the following prescription.

For any given positive integer N, we list out all its divisors:

$$d_1, \quad d_2, \quad d_3, \quad d_4, \quad \ldots.$$

(The list must include the divisors 1 and N too.) Let us call this the "d-list". Now, we find out the *number* of divisors that each of these divisors has. Let the divisor d_i have a_i divisors. We now have a new list, the "a-list":

$$a_1, \quad a_2, \quad a_3, \quad a_4, \quad \ldots.$$

This list has the feature we have in mind: the sum of the cubes of the numbers is equal to the square of their sum. We give two examples to show how the prescription works.

(a) Let $N = 6$. The divisors of 6 are 1, 2, 3 and 6. To find the a-list, note that

1 has 1 divisor (namely, 1 itself);
2 has 2 divisors (namely, 1 and 2);
3 has 2 divisors (namely, 1 and 3); and
6 has 4 divisors (namely, 1, 2, 3 and 6).

Thus, the a-list is 1, 2, 2, 4. Now, observe that

$$1^3 + 2^3 + 2^3 + 4^3 = 81 = (1 + 2 + 2 + 4)^2.$$

(b) Let $N = 30$. The divisors of 30 are 1, 2, 3, 5, 6, 10, 15 and 30, and these numbers have

$$1, \quad 2, \quad 2, \quad 2, \quad 4, \quad 4, \quad 4, \quad 8$$

divisors respectively. Now, note that the sum of these numbers is 27, and the sum of their cubes is 729 or 27^2.

By choosing N to be a power of 2 (or, more generally, a power of any prime number), we recover the result with which we started this discussion. For example: Let $N = 2^4 = 16$. The divisors of 16 are 1, 2, 4, 8 and 16, and these numbers have (respectively) 1, 2, 3, 4 and 5 divisors; and of course,

$$1^3 + 2^3 + 3^3 + 4^3 + 5^3 = (1 + 2 + 3 + 4 + 5)^2.$$

8.2 Last digit

Listing out the cubes in normal decimal notation,

$$1, \quad 8, \quad 27, \quad 64, \quad 125, \quad 216, \quad 343, \quad 512, \quad 1000, \quad \ldots$$

we find that any digit from 0 to 9 may occur as the units digit. (For the squares, the units digits formed a restricted set.) This simple observation serves as the basis for a simple but entertaining trick which, from the author's experience, can appear most impressive in a classroom setting.

We announce to the audience that we can instantly find the cube root of any perfect cube less than one million. That is, given the value of N^3 we can instantly find N, provided that N lies between 1 and 100. We need to memorize two lists. The first is the list of cubes less than 1000. There are only just 9 such cubes (as listed above), and they are easy to remember.

The second list is the following:

units digit of n	0	1	2	3	4	5	6	7	8	9
units digit of n^3	0	1	8	7	4	5	6	3	2	9

which gives the units digit of each cube, matched against the units digit of the original number. Observe that the table has several symmetries: (a) 2^3 ends in 8, and 8^3 ends in 2; (b) 3^3 ends in 7, and 7^3 ends in 3; and so on. These make the table easy to memorize.

Now for how the trick works. Given any cube, say $N^3 = 300763$, we first note its units digit, 3 and deduce, using the above table, that the units digit of N is 7. Next, we discard its last three digits and we are left with the number 300, which lies between the cubes 216 and 343, that is, between the cubes of 6 and 7. We take the smaller of these numbers (that is, 6), and deduce that the tens digit of N is 6. That is, $N = 67$.

If we were given the number 148877, the '7' in the units place tells us that the units digit of N is 3; and since 148 lies between 5^3 and 6^3, we deduce that its tens digit is 5; so $N = 53$.

Likewise, given the number 405224, the '4' in the units place tells us that the units digit of N is 4; and since 405 lies between 7^3 and 8^3, the cube root must be 74.

Now, try out the trick on your friends!

★ ★ ★

It is interesting to ask whether the above trick can be done for *fifth powers*; the answer is: yes! The fifth powers less than 10^5 are:

1, 32, 243, 1024, 3125, 7776, 16807, 32768, 59049.

We quickly observe that for each value of N, the units digit of N^5 is the same as that of N. So to recover N from N^5 we do the following: (a) note its units digit; this will be the units digit of N and (b) delete its *last five digits*. From the leftover number we find the tens digit of N, using the list shown above. (The list should be memorized.)

For example, given that $N^5 = 1048576$ we first deduce that the units digit of N is 6. Next, deleting the last five digits we are left with the number 10, which lies between 1 and 32 (the fifth powers of 1 and 2). Therefore, the tens digit of N is 1, and so $N = 16$.

If we were given that $N^5 = 69343957$, then we deduce first that the units digit of N is 7. Discarding the last five digits we are left with the number 693, which lies between 243 and 1024 (the fifth powers of 3 and 4). Therefore, the tens digit of N is 3, and so $N = 37$.

Similarly, from $N^5 = 916132832$ we deduce that $N = 62$, from $N^5 = 2706784157$ we deduce that $N = 77$, and from $N^5 = 7339040224$ we obtain $N = 94$.

Now try *that* on your friends. The results should be most interesting!

8.3 Coin changing with cubes

Earlier, we saw that every positive integer is the sum of no more than four squares. Can a similar statement be made about cubes? The answer is *yes*, but the complexity of the matter turns out to be substantially greater than before. We actually need to ask two questions:

(a) Is there a number k, such that every integer is the sum of k or fewer positive cubes?

(b) Is there a number $k' < k$, such that every sufficiently large integer is the sum of k' or fewer positive cubes? (We explain below what "sufficiently large integer" means.)

The numbers k and k' are referred to by mathematicians as $g(3)$ and $G(3)$, respectively.

In the case of the squares, the word "positive" was not necessary, as the squares are all non-negative. Of greater consequence is the fact that the two questions collapse into one; for (a) the answer is 4, and for (b) it is that infinitely many integers require 4 squares (we cannot do with less). Indeed, the numbers belonging to the AP

$$7, \quad 15, \quad 23, \quad 31, \quad 39, \quad 47, \quad 55, \quad 64, \quad \ldots$$

all require 4 squares, and there are infinitely many numbers on this list. So it cannot be the case that from some point on, 3 or fewer than 3 squares will suffice. In mathematical notation, we write $g(2) = G(2) = 4$.

In the case of the cubes, however, the two questions do not have the same answer; but to show this is very difficult! Observe that 7 requires 7 cubes ($7 = 7 \times 1^3$), and we obviously cannot do with less, because 7 is less than 2^3, implying that we can use only 1s. Similarly, consider the number 23; since $23 < 3^3$, we can use only the cubes of 1 and 2, and since $23 < 3 \times 2^3$, the best we can do is

$$23 = (2 \times 2^3) + (7 \times 1^3).$$

So 23 requires as many as 9 cubes. Are there numbers which need more than 9 cubes? The answer is *no*; but, as indicated above, it is hard to prove this. So we have a theorem here:

> *Every integer is a sum of* 9 *or fewer positive cubes. Moreover, there are numbers that cannot be represented by fewer than* 9 *positive cubes. In other words,* $g(3) = 9$.

The motivation for question (b) is as follows. We have already noted that 23 requires 9 cubes. If we run through the numbers following 23, one by one, we find that most of them do not require that many cubes. Indeed, the next number which requires 9 cubes is 239:

$$239 = (2 \times 4^3) + (4 \times 3^3) + (3 \times 1^3),$$

and we find that no number from 240 till 40000 requires that many! Indeed, the available evidence suggests that the only numbers requiring 9 cubes are 23 and 239. So, it seems likely that from some point on (this is what the phrase "every sufficiently large number" means) we can make do with 8 (or perhaps fewer) cubes. Actually, more can be said: it is found that between 240 and 40000, the only number that requires 8 cubes is 454:

$$454 = (1 \times 7^3) + (4 \times 3^3) + (3 \times 1^3).$$

So is it possible that from 455 onwards, every number is a sum of 7 or fewer cubes? Examining the tables which have been prepared with great care by mathematicians, we find that there are very few numbers which require as many as 7 cubes; after 8042 we do not seem to require more than 6 cubes! The number 8042 itself requires 7 cubes:

$$8042 = (1 \times 19^3) + (2 \times 7^3) + (2 \times 6^3) + (1 \times 4^3) + (1 \times 1^3).$$

The evidence thus suggests that from 8043 onwards 6 cubes will suffice; in other words, that $G(3) = 6$, or less. And it is possible that from some point on, even fewer than 6 cubes will suffice! Will this progression ever come to an end?

Arguing from the reverse end, we can show fairly easily that $G(3)$ cannot be less than 4; so the progression does come to an end! This is because numbers belonging to the AP

$$4, \quad 13, \quad 22, \quad 31, \quad 40, \quad 49, \quad 58, \quad 67, \quad 76, \quad \ldots$$

cannot be represented by fewer than 4 cubes. To see why this is so, it is enough to note that these numbers are all of the form $4 + 9k$, and that all cubes are of one of the forms

$$9k, \quad -1 + 9k, \quad 1 + 9k.$$

(To understand why, simply enumerate the cubes from 1^3 to 9^3 and examine their remainders when divided by 9.) So, no 3 such numbers can yield a number of the form $4 + 9k$.

Thus, $G(3)$ is equal to 4, 5, 6, 7, 8, or 9; but which one? We know that $G(3)$ must be one of the numbers 4, 5, 6 or 7 (the numbers 8 and 9 can be ruled out with some effort), but beyond this no one knows very much more!

8.4 More sums of cubes

In the section on "Squares" we had asked about instances when the sum of two squares is another square. This question may be asked of cubes too: *Are there instances when the sum of two cubes is yet another cube?* The reader who is eagerly looking forward to seeing such sets of cubes is doomed to disappointment, because there are no such sets. That is, *the relation* $a^3 + b^3 = c^3$ *cannot hold for positive integers* a, b, c.

This claim is, of course, a special case of one of the most famous theorems (perhaps we should add the word "notorious") in all of mathematics – Fermat's "Last Theorem", according to which the equation

$$a^n + b^n = c^n$$

cannot hold in positive integers a, b, c if n is an integer greater than 2. The theorem was stated by Fermat in the margin of a book, with the *claim* of a proof but no actual proof. It was proved only in 1993–95 by Andrew Wiles, using extremely difficult techniques that were developed only in this century. The special case $n = 4$ is easy to handle using the idea of infinite descent (we made a mention of this proof technique earlier on, in the section on "Squares"). The case $n = 3$ is already quite difficult; the first proof of this case is due to Gauss. There is no question of our giving the proof here, unfortunately – it is much too intricate. Wiles's proof is, sadly, still less accessible! However, using the techniques developed by him, more results have been obtained. For instance, it has been shown that there are no instances of three distinct, positive cubes in AP.

Since the sum of two cubes will never yield another cube, we shall ask a different question. Are there solutions in the positive integers a, b, c, d to the equation

$$a^3 + b^3 = c^3 + d^3?$$

An equivalent question is the following: *Are there positive integers which can be written as a sum of two cubes in more than one way?* The answer is *yes*, and Ramanujan's "taxicab number", 1729, provides the smallest (and most famous) example of such a number:[1]

[1] There is a very well-known incident associated with this number which accounts for the use of the phrase "taxicab". On one occasion, Hardy was visiting the ailing Ramanujan in a sanatorium. To make conversation, perhaps, he remarked that the taxicab in which he had travelled had the number 1729, and that it did not seem "too interesting". At this Ramanujan replied promptly that on the contrary it was an extremely interesting number, being the smallest number that is the sum of two cubes in two different ways. Hardy narrated later that Ramanujan seemed to regard the positive integers as friends, and knew their idiosyncrasies, much as we do of our friends.

$$1729 = 9^3 + 10^3 = 1^3 + 12^3.$$

Other instances are given by the numbers 4104 and 39312. We have,

$$\begin{aligned} 4104 &= 9^3 + 15^3 = 2^3 + 16^3, \\ 39312 &= 2^3 + 34^3 = 15^3 + 33^3. \end{aligned}$$

If we allow the use of negative numbers, we also have

$$\begin{aligned} 81 &= 3^3 + 4^3 = (-5)^3 + 6^3, \\ 217 &= 1^3 + 6^3 = (-8)^3 + 9^3. \end{aligned}$$

The first of these relations may be expressed in a more pleasing form as follows:[2]

$$3^3 + 4^3 + 5^3 = 6^3.$$

The least positive integer which admits of *three* different representations is 4104. Two representations are as given above, and the third one is

$$4104 = (-12)^3 + 18^3.$$

However, this method uses negative numbers. If we insist that only positive numbers are used, then the least integer that allows three representations is 87539319. We have,

$$\begin{aligned} 87539319 &= 436^3 + 167^3 \\ &= 423^3 + 228^3 = 414^3 + 255^3. \end{aligned}$$

As you may have guessed, finding such instances is hard work!

* * *

Here is another curiosity. We may ask whether there exist any numbers which equal the sums of the cubes of their digits (when written in normal decimal notation). It turns out that there are precisely five such numbers: 1, 153, 370, 371 and 407. For the four numbers greater than 1, we check this claim as follows:

$$\begin{aligned} 153 &= 1^3 + 5^3 + 3^3, \\ 370 &= 3^3 + 7^3 + 0^3, \\ 371 &= 3^3 + 7^3 + 1^3, \\ 407 &= 4^3 + 0^3 + 7^3. \end{aligned}$$

---

[2] Examining the relations $3^2 + 4^2 = 5^2$ and $3^3 + 4^3 + 5^3 = 6^3$, we may wonder whether the numbers $3^4 + 4^4 + 5^4 + 6^4$ and 7^4 are equal; unfortunately, they are not.

Here, the cubes score above the squares, because there are no numbers which equal the sums of the squares of their digits (when written in decimal notation). This is not hard to prove. On the other hand, there do exist numbers which equal the sums of the 4th powers of their digits (as earlier, when written in decimal notation). For instance, 8208 is one such number:

$$8208 = 8^4 + 2^4 + 0^4 + 8^4.$$

There are two other such 4-digit numbers, namely 1634 and 9474. For 5th powers too we find such numbers, for instance 4150 (we have $4150 = 4^5 + 1^5 + 5^5 + 0^5$).

⋆ ⋆ ⋆

There is a curious way in which such numbers can be found, and this is in itself worth an investigation. Starting with any number written in normal decimal notation, we find the sum of the cubes of its digits, then the sum of the cubes of the digits of the resulting number, then the sum of the cubes of the digits of *that* number, and so on. After a few steps, we discover to our surprise that the numbers stay fixed, and the number we have before us is one of the numbers listed above (namely, 1, 153, 370, 371 or 407). For example, if we start with the number 998, we get:

$$998, \quad 1970, \quad 1073, \quad 371, \quad 371, \quad 371, \quad \ldots;$$

and if we start with 997, we get:

$$997, \quad 1801, \quad 514, \quad 190, \quad 730, \quad 370, \quad 370, \quad \ldots;$$

and so on. The examples for 4th powers and 5th powers may also be found in a similar manner.

8.5 One cube twice another?

Earlier, we had showed that there are no instances of two squares where one is twice the other. The same proof holds true for cubes too (practically verbatim). So, the equation

$$a^3 = 2b^3$$

does not admit solutions in positive integers.

We may be tempted, as we were earlier, to ask for "near-miss" solutions; that is, solutions in positive integers to the two equations

$$a^3 - 2b^3 = 1, \qquad a^3 - 2b^3 = -1.$$

Observe that these are *cubic equations*; the equations we studied earlier were *quadratic* in nature. We quickly find an important difference between the two kinds of equations. We have earlier encountered quadratic equations with infinitely many integral solutions; e.g., the equation $a^2 - 2b^2 = 1$. It can be shown that *a polynomial equation of degree 3 or more cannot have infinitely many integer solutions.* This was conjectured in the early 1900s by a mathematician named L J Mordell, but it was proved only in the 1980s (by Gerd Faltings). The cubic equations listed above provide good examples of the Mordell–Faltings theorem: the equation $a^3 - 2b^3 = 1$ has precisely one integer solution, namely $a = 1$, $b = 0$; and the equation $a^3 - 2b^3 = -1$ too has precisely one integer solution, namely $a = 1$, $b = 1$. As with so many statements made in this chapter, however, the proofs of these statements are difficult.

⋆ ⋆ ⋆

The question posed above arises when we analyze the following problem. *The cube 8 and the square 9 differ by 1. Are there other such pairs of numbers?* Interestingly, the answer is: no. In other words, the only solution in positive integers to the equation

$$a^3 - b^2 = \pm 1$$

is $a = 2$, $b = 3$. We briefly show how the two problems are connected. To simplify the discussion, we shall drop the word "positive" and ask only for integral solutions to the equation.

Consider first the equation $a^3 - b^2 = -1$, which we write as $a^3 = (b-1)(b+1)$. If b is even, then $b-1$ and $b+1$ are consecutive odd numbers and therefore, share no common factor. Since their product is a^3, a cube, each of them is a cube. So we have two cubes which differ by 2, and the only such pair of cubes is the pair -1 and 1. This means that $b+1 = 1$ and $b-1 = -1$, giving $b = 0$; therefore, $a = -1$ and $b = 0$ is the only solution in which b is even.

If b is odd, then $b-1$ and $b+1$ are consecutive even numbers, so their gcd is 2. Clearly, one of the numbers $b-1, b+1$ is a multiple of 4, and the other is not. If $b-1$ is a multiple of 4, then we write

$$\left(\frac{b-1}{4}\right)\cdot\left(\frac{b+1}{2}\right)=\left(\frac{a}{2}\right)^3.$$

In this case, the numbers $(b-1)/4$ and $(b+1)/2$ share no common factor (the common factor 2 has been divided out) and, as their product is a cube, each of them must be a cube; say $(b-1)/4=c^3$ and $(b+1)/2=d^3$. We now get,

$$4c^3+1=b,\quad 2d^3-1=b,\quad \therefore\ 4c^3+1=2d^3-1,$$

and so $d^3-2c^3=1$. The only solution to this equation is $d=1$, $c=0$, so we get $a=0$ and $b=1$.

The other possibility is that $b+1$ is a multiple of 4. In this case, we write

$$\left(\frac{b-1}{2}\right)\cdot\left(\frac{b+1}{4}\right)=\left(\frac{a}{2}\right)^3.$$

We deduce, as we did earlier, that the numbers $(b-1)/2$ and $(b+1)/4$ are cubes, say $(b-1)/2=c^3$ and $(b+1)/4=d^3$. This yields:

$$2c^3+1=b,\quad 4d^3-1=d,\quad \therefore\ 2c^3+1=4d^3-1,$$

and so $c^3-2d^3=-1$, whose only solution is $c=1$, $d=1$. This gives the result $a=2$ and $b=3$, and we have found the only "interesting" solution to $a^3-b^2=-1$.

The equation $a^3-b^2=1$ can be handled in a similar manner, but the details are awkward, and we shall not pursue the matter any further here.

Exercises

8.5.1 Show that all cubes are of the form $9k$ or $9k\pm1$.

8.5.2 Show that if n is not divisible by 13, then n^3 is of one of the forms $13k\pm1$ or $13k\pm5$.

8.5.3 Show that if n is not divisible by 19, then n^3 is of one of the forms $19k\pm1$; $19k\pm7$ or $19k\pm8$.

8.5.4 Verify that the numbers from 1 to 100 are all sums of 9 or fewer cubes.

8.5.5 Find a positive integer N such that $N/2$ is a square, $N/3$ is a cube and $N/5$ is a fifth power.

Chapter 9

A Curious Procedure

In this chapter, we shall briefly describe a curious procedure that generates many sequences of interest. The reader may be hard pressed to explain why the procedure accomplishes what it does!

9.1 Details of the procedure

The procedure involves cumulatively adding up the elements of the sequence of natural numbers

$$A = \langle 1,\ 2,\ 3,\ 4,\ 5,\ 6,\ 7,\ 8,\ 9,\ 10,\ \ldots \rangle$$

after first selectively deleting certain terms of the sequence. The deletions are done as follows.

We choose a positive integer k and delete every kth term in the sequence. (So the first number to be deleted is k, the next is $2k$, and so on.) Having done the deletion, we obtain a new sequence which we call B:

$$B = \langle b_1,\ b_2,\ b_3,\ b_4,\ b_5,\ b_5,\ b_6,\ \ldots \rangle.$$

We now compute the *partial sums* of B. That is, we compute the sequence C whose elements $c_1, c_2, c_3, \ldots,$ are given by

$$c_1 = b_1, \quad c_2 = b_1 + b_2, \quad c_3 = b_1 + b_2 + b_3,$$

and so on. Now, we have the sequence C of partial sums of B.

We then delete every $(k - 1)$ th term of C, call the resulting sequence D, and compute *its* partial sums; we get another sequence, E. Next, we delete every $(k - 2)$ th term of E, call the resulting sequence F, and find the partial sums of F; we call the resulting sequence G; and so on. After each step the gap between the deleted members decreases by 1, until finally the gap becomes 2. The partial sums computed at this point usually present a major surprise.

Symbolically, we may write

$$A \xrightarrow{\Delta_n} B \xrightarrow{\Sigma} C,$$

to denote the two operations: sequence B is obtained from sequence A by deleting every n^{th} term (we represent the deletion operation by the Greek letter Δ, pronounced "Delta"), and C is obtained from B by cumulative addition (we represent the operation by the Greek letter Σ, which is pronounced "Sigma" and generally represents addition). So, the sequence of operations may be depicted as shown below:

$$\begin{array}{ccccc} A & \xrightarrow{\Delta_n} & B & \xrightarrow{\Sigma} & C, \\ C & \xrightarrow{\Delta_{n-1}} & D & \xrightarrow{\Sigma} & E, \\ E & \xrightarrow{\Delta_{n-2}} & F & \xrightarrow{\Sigma} & G, \end{array}$$

and so on. The last line is, writing P, Q, R for the sequences involved,

$$P \xrightarrow{\Delta_2} Q \xrightarrow{\Sigma} R,$$

and our claim is that the sequence R possesses some unexpected features of interest.

The reader will naturally want to see the procedure in action. We give some examples below.

Examples

9.1.1 The case $k = 2$

Here $A = \langle 1, 2, 3, 4, 5, 6, 7, \ldots \rangle$; after deleting every 2nd term, we get the sequence

$$1,\ 3,\ 5,\ 7,\ 9,\ 11,\ \ldots,$$

and the partial sums of these numbers are

$$1,\ 4,\ 9,\ 16,\ 25,\ 36,\ \ldots.$$

We have thus obtained the sequence of squares.

9.1.2 The case $k = 3$

We start with the same sequence, A, and delete every 3rd term; the result is the sequence

$$1,\ 2,\ 4,\ 5,\ 7,\ 8,\ 10,\ 11,\ 13,\ 14,\ \ldots.$$

The partial sums of this sequence are

$$1,\ 3,\ 7,\ 12,\ 19,\ 27,\ 37,\ 48,\ 61,\ 75,\ \ldots.$$

We now delete every 2nd term of this sequence, and we are left with

$$1,\ 7,\ 19,\ 37,\ 61,\ \ldots,$$

and the partial sums of this sequence are

$$1,\ 8,\ 27,\ 64,\ 125,\ \ldots;$$

this is just the sequence of cubes!

9.1.3 The case $k = 4$

For $k = 4$ and higher values, it is much more convenient to work directly on an array as shown below. We obtain the sequence of fourth powers!

A	1	2	3	4	5	6	7	8	9	10	11	12	13
B	1	2	3		5	6	7		9	10	11		13
C	1	3	6		11	17	24		33	43	54		67
D	1	3			11	17			33	43			67
E	1	4			15	32			65	108			175
F	1				15				65				175
G	1				16				81				256

9.1.4 The case $k = 5$

Here, we proceed as shown below. Observe the fifth powers in the last line!

A	1	2	3	4	5	6	7	8	9	10	11	12	13	14
B	1	2	3	4		6	7	8	9		11	12	13	14
C	1	3	6	10		16	23	31	40		51	63	76	90
D	1	3	6			16	23	31			51	63	76	
E	1	4	10			26	49	80			131	194	270	
F	1	4				26	49				131	194		
G	1	5				31	80				211	405		
H	1					31					211			
I	1					32					243			

⋆⋆⋆

The evidence is unmistakable, and we hypothesize that:

> *If from the sequence 1, 2, 3, . . . we delete every n^{th} term, then every $(n-1)^{\text{st}}$ term, then every $(n-2)^{\text{nd}}$ term, . . ., and lastly every 2nd term, computing the partial sums after each round of deletions, then the final sequence of partial sums is the sequence of nth powers.*

How does one explain this curious phenomenon?

It is possible to generalize the procedure somewhat and obtain as a result yet other sequences; please see reference [3] in Appendix C.

Chapter 10

Powers of 2 and 3

In this section we explore a few of the many properties of the powers of 2:

$$1,\ 2,\ 4,\ 8,\ 16,\ 32,\ 64,\ 128,\ \ldots,$$

and the powers of 3:

$$1,\ 3,\ 9,\ 27,\ 81,\ 243,\ 729,\ 2187,\ \ldots.$$

Observe that we start both sequences with 1 ($= 2^0 = 3^0$) and not 2 and 3, respectively. However, on some occasions, we shall find it more convenient to start with 2^1 and 3^1 respectively.

One of the simplest patterns to be found in the two sequences has to do with their cumulative partial sums. In the case of powers of 2, the partial sums obtained are

$$\begin{aligned} 1 &= 1, \\ 1+2 &= 3, \\ 1+2+4 &= 7, \\ 1+2+4+8 &= 15, \end{aligned}$$

.... so the sequence of partial sums is 1, 3, 7, 15, 31, 63, ...; each number is 1 short of the *next* power of 2! Stated algebraically, this observation reads as follows: for all positive integers n,

$$1 + 2 + 2^2 + 2^3 + \cdots + 2^n = 2^{n+1} - 1.$$

Once expressed concretely, it is easy to prove. (This often happens when we translate statements into algebra. "Algebra is kind – she always gives us more than we ask for".) We write A for the sum on the left side:

$$A = 1 + 2 + 2^2 + 2^3 + \cdots + 2^n.$$

Multiplying the entire equation by 2, we get

$$2A = 2 + 2^2 + 2^3 + \cdots + 2^n + 2^{n+1}.$$

Nice things happen when we subtract the 1st equation from the 2nd; almost all the terms cancel out, and we are left with

$$A = 2^{n+1} - 1,$$

which is exactly what we wanted. So the observed relation is a reality.

A similar relation is seen with the powers of 3:

$$\begin{aligned} 1 &= 1 = \tfrac{1}{2}(3-1), \\ 1+3 &= 4 = \tfrac{1}{2}(9-1), \\ 1+3+9 &= 13 = \tfrac{1}{2}(27-1), \\ 1+3+9+27 &= 40 = \tfrac{1}{2}(81-1), \end{aligned}$$

.... Here, the relation is: for all positive integers n,

$$1 + 3 + 3^2 + 3^3 + \cdots + 3^n = \tfrac{1}{2}\left(3^{n+1} - 1\right).$$

As earlier, once so expressed it is easy to prove; exactly the same strategy works, except that this time we multiply by 3 and not 2. The details are left to the reader.

10.1 Last digit

Another simple pattern to be seen in the powers of 2 lies in the sequence of last digits, when the numbers are written in normal decimal notation:

$$2,\ 4,\ 8,\ 6,\ 2,\ 4,\ 8,\ 6,\ 2,\ 4,\ 8,\ 6,\ \ldots.$$

Observe that the string 2, 4, 8, 6 repeats over and over again. Indeed it must, for once we reach 6, the next number in the sequence has 2 for its units digit (because $2 \times 6 = 12$, whose units digit is 2). Next, 2 is followed by 4 (because $2 \times 2 = 4$), which in turn is followed by 8, and 8 is followed by 6 once again, thus completing the cycle. So the logic behind the repetition is easy to see.

It is less obvious but true that the sequence of *last two digits* also repeats, and for exactly the same reason. However, the cycle length is longer. Writing 2 as 02, 4 as 04, and so on, the sequence of last two digits goes thus:

02, 04, 08, 16, 32, 64, 28, 56, 12, 24, 48, 96, 92, 84,
68, 36, 72, 44, 88, 76, 52, 04, 08, 16, 32, ...

The combination 02 does not reappear after its occurrence at the start, and the string

$$\langle 04,\ 08,\ 16,\ 32,\ \ldots,\ 88,\ 76,\ 52\rangle$$

repeats endlessly. The repeating portion is therefore a string with 20 numbers. To see why it repeats, observe that once 52 is reached, the numbers to follow are 04, 08, 16, and so on; the same numbers reappear.

It will not come as a surprise that the sequence of last three digits also repeats:

002, 004, 008, 016, 032, 064, 128, 256, 512, 024, 048,
096, 192, 384, 768, 536, 072, 144, 288, 576, 152, 304,
608, 216, 432, 864, 728, 456, 912, 824, 648, 296, 592,
184, 368, 736, 472, 944, 888, 776, 552, 104, 208, 416,
832, 664, 328, 656, 312, 624, 248, 496, 992, 984, 968,
936, 872, 744, 488, 976, 952, 904, 808, 616, 232, 464,
928, 856, 712, 424, 848, 696, 392, 784, 568, 136, 272,
544, 088, 176, 352, 704, 408, 816, 632, 264, 528, 056,
112, 224, 448, 896, 792, 584, 168, 336, 672, 344, 688,
376, 752, 504, 008, 016, ...

Here, the first two entries (002, 004) appear only at the start; they never occur again. So the repeating string contains the following 100 numbers (the dots represent 93 numbers): ⟨008, 016, 032, 064, ..., 376, 752, 504⟩.

⋆ ⋆ ⋆

More generally, if we consider only the last k digits of the powers of 2 (written in normal decimal notation), k being any given positive integer, then the sequence repeats no matter what k is, though some initial terms may have to be thrown away. If we take $k = 4$ (here we focus attention on the last four digits), then the first three entries (0002, 0004, 0008) do not reappear after their initial occurrence, and the repeating portion is a string containing 500 numbers:

$$\langle 0016, 0032, 0064, 0128, 0256, 0512, 1024, 2048, 4096, 8192, 6384, 2768, \ldots, 8752, 7504, 5008\rangle.$$

The alert reader may have noticed that we have hit upon a pattern within a pattern! If we write r_k for the length of the portion that repeats when we consider only the last k digits, then we find the following values for r_k:

k	1	2	3	4	…
r_k	4	20	100	500	…

The pattern is not hard to spot—each successive value of r_k is precisely 5 times the preceding value!

⋆⋆⋆

It is remarkable that virtually the same statements can be made about the powers of 3 as about the powers of 2. As earlier, the units digit cycles through a string containing 4 numbers:

$$1,\ 3,\ 9,\ 7,\ 1,\ 3,\ 9,\ 7,\ 1,\ \ldots.$$

The repeating string is $\langle 1,3,9,7\rangle$.

If we consider the last two digits, the repeating portion is a string containing 20 numbers:

$$01, 03, 09, 27, 81, 43, 29, 87, 61, 83, 49, 47, 41, 23, 69, 07, 21, 63, 89, 67, 01, 03, 09, \ldots.$$

The repeating portion here is the string

$$\langle 01,\ 03,\ 09,\ \ldots,\ 63,\ 89,\ 67\rangle.$$

One difference that we immediately notice is that the repeating portion includes the starting term (01 in this case); the initial terms are not lost, as had happened in the case of the powers of 2.

If we consider the last three digits, then the repeating portion turns out to be a string containing 100 numbers:

$$001, 003, 009, 027, 081, 243, 729, 187, 561, 683, 049, 147, 441, 323, 969, 907, 721, 163, 489, 467, 401, 203,$$

609, 827, 481, 443, 329, 987, 961, 883, 649, 947, 841,
523, 569, 707, 121, 363, 089, 267, 801, 403, 209, 627,
881, 643, 929, 787, 361, 083, 249, 747, 241, 723, 169,
507, 521, 563, 689, 067, 201, 603, 809, 427, 281, 843,
529, 587, 761, 283, 849, 547, 641, 923, 769, 307, 921,
763, 289, 867, 601, 803, 409, 227, 681, 043, 129, 387,
161, 483, 449, 347, 041, 123, 369, 107, 321, 963, 889,
667, 001, 003, 009, ...

The repeating portion here is the string

$$\langle 001,\ 003,\ 009,\ \ldots,\ 963,\ 889,\ 667\rangle.$$

Notice that the initial terms (001, 003) are part of the cycle.

In the case of the last four digits the repeating portion contains 500 digits, and it starts and ends as shown below:

0001, 0003, 0009, 0027, 0081, 0243, 0729, 2187, 6561,
9683, 7147, . . . 4321, 2963, 8889, 6667.

The initial terms are once again part of the cycle.

The reader may wonder whether the same cycle lengths occur for other powers, say for the powers of 5, 6, 7, ... The answer is: *no*. For example,

- for the powers of 5, the units digit is always 5, the last two digits are always 25, the last three digits are 125 and 625 in alternation (so the cycle lengths are 1, 1 and 2, respectively), and so on;
- for the powers of 6, the units digit is always 6, the last two digits cycle through the string $\langle 36, 16, 96, 76, 56\rangle$, the last three digits cycle through the string

  $$\langle 216, 296, \ldots, 256, 536\rangle$$

  with length 25, and so on.

⋆ ⋆ ⋆

Here is a curious fact about the powers of 3.

Consider first the powers of 2. Since 2 is even, the units digit of a power of 2 is even and can only take the values 2, 4, 6 or 8 (it can never be 0, as a power of 2 is not divisible by 5). The tens digit, on the other hand, can take any value. Scanning through a list of powers of 2, we find all the possible digits occurring in the tens places (0 is the tens digit of 2^3 and also of 2^{23}; 1 is the tens digit of 2^4; 2 is the tens digit of 2^{10}; and so on). Nothing unusual to report here!

Now, consider the powers of 3. Since 3 is odd, the units digit of a power of 3 is odd and can only take the values 1, 3, 7 or 9 (it can never be 5, because a power of 3 is not divisible by 5). What about the tens digit? Surprise!—we find that *the tens digit is invariably even!* The list displayed below brings this out (the tens digit has been highlighted in each case; in the case of 3^0, 3^1 and 3^2, the tens digit may be taken as 0, an even number):

> 01, 03, 09, 27, 81, 243, 729, 2187, 6561, 19683,
> 59049, 177147, 531441, ...

What a curious phenomenon! How do we account for it? Here is one way.

The possible units digits of a power of 3 are 1, 3, 7 and 9. Suppose that the tens digit of a particular power of 3, say 3^k, is even; call it $2a$. Then the number represented by the last two digits of 3^k is one of the numbers

$$20a+1,\ 20a+3,\ 20a+7,\ 20a+9.$$

Now, consider the next power of 3, namely 3^{k+1}. The number representing *its* last two digits will be determined by the last two digits of 3^k. Multiplying each of the above numbers by 3, we obtain the numbers

$$60k+3,\ 60k+9,\ 60k+21,\ 60k+27.$$

In the last two numbers, a '2' gets carried into the tens place from the units place (from the products 21 and 27 respectively), whereas in the first two numbers there is no carry-over (the carry-over is 0). *So the carry-over in each case is an even number.*

Also, the result of multiplying the tens digit by 3 yields an even digit, because the original digit itself is even. Adding the carry-over, we once again obtain an even number, because even plus even is even. So the tens digit of 3^{k+1} also is even!

Thus, an even digit in the tens place of 3^k results in an even digit in the tens place of 3^{k+1}; and this, in turn, results in an even digit in the tens place of 3^{k+2}; and so on, for ever and ever! This means that if the tens digit of any one power of 3 is even, then this will be true for *every* higher power of 3.

But the property in question *is* is true for 3^0 and 3^1. Therefore, it is true for every power of 3!

10.2 Number of digits

Let the number 2^n be written in normal decimal notation, and let $\#(2^n)$ be its number of digits.

Example (a) $2^1 = 1$, so $\#(2^1) = 1$; (b) $2^3 = 16$, so $\#(2^4) = 2$; (c) $2^{10} = 1024$, so $\#(2^{10}) = 3$.

Computing $\#(2^n)$ for large n can prove cumbersome, but it is easy if we have at our disposal a computer algebra package such as Mathematica or Derive. Numerical experimentation yields the following data, which we have summarized below (see Table B.10.1).

Table B.10.1 *Powers of 2 having a given number of digits*

k	Values of n for which $\#(2^n) = k$
1	1, 2, 3
2	4, 5, 6
3	7, 8, 9
4	10,11, 12, 13
5	14, 15, 16
6	17, 18, 19
7	20, 21, 22, 23
8	24, 25, 26
9	27, 28, 29
10	30, 31, 32, 33
11	34, 35, 36
12	37, 38, 39

The results are striking – it appears that for each value of k there are either 3 or 4 values of n for which 2^n has k digits. Is this a genuine pattern? Yes, and it is not hard to show why.

Consider the very first power of 2 which has k digits (k being a given positive integer). Call this number N. (For example, if $k = 3$ then $N = 128$, and if $k = 4$ then $N = 1024$.)

Clearly, we have the following inequality:

$$\underbrace{1000\ldots000}_{(k-1)\ \text{zeroes}} < N < \underbrace{2000\ldots000}_{(k-1)\ \text{zeroes}}.$$

Repeated multiplication by 2 yields the following:

$$\begin{array}{rcccl} 1000\ldots000 & < & N & < & 2000\ldots000 \\ 2000\ldots000 & < & 2N & < & 4000\ldots000 \\ 4000\ldots000 & < & 4N & < & 8000\ldots000 \\ 8000\ldots000 & < & 8N & < & 16000\ldots000 \\ 16000\ldots000 & < & 16N & < & 32000\ldots000 \end{array}$$

So N, $2N$, $4N$ all have k digits; $8N$ *may* have the same number of digits, and $16N$ certainly has $(k+1)$ digits. Therefore, the number of powers of 2 with k digits is either 3 or 4.

To find a formula for $\#(2^n)$, we display the results in the form of a table.

n	10	20	30	40	50	60	70	80	90	100
$\#(2^n)$	4	7	10	13	16	19	22	25	28	31

The data may also be shown in the form of a graph, as below (Figure B.10.1).

The graph wiggles a lot, but its overall straightness is quite striking. The table reveals that $\#(2^n)$ is close to (but a bit larger than) $0.3n$. Detailed analysis shows that an excellent formula is

$$\#(2^n) = 1 + \text{integer part of } 0.30103 \times n.$$

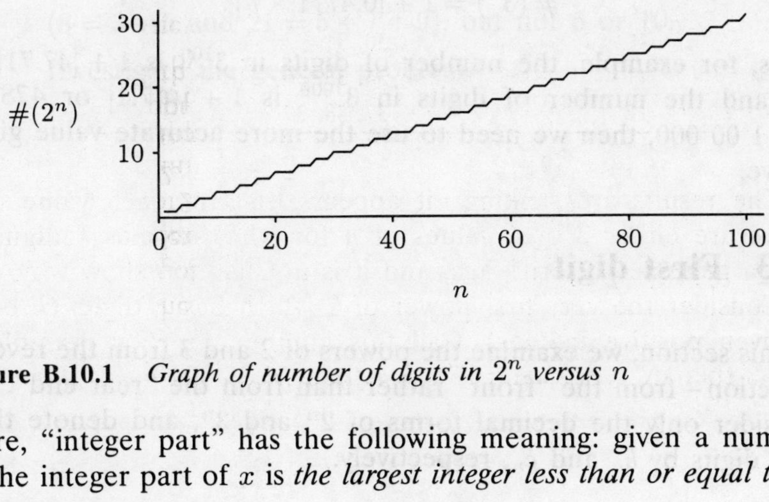

Figure B.10.1 *Graph of number of digits in 2^n versus n*

Here, "integer part" has the following meaning: given a number x, the integer part of x is *the largest integer less than or equal to x.*

Generally, we write $[x]$ to denote the integer part of x. For example, $[3.8] = 3$, $[4.1] = 4$, and so on. To show how the formula is used, consider the sample values $n = 50$ and $n = 100$. We have,

$$\begin{aligned} \#(2^{50}) &= 1 + [0.30103 \times 50] = 1 + 15 = 16, \\ \#(2^{100}) &= 1 + [0.30103 \times 100] = 1 + 30 = 31, \end{aligned}$$

and these are correct. Handling large values of n now becomes easy. For example, we may deduce that 2^{1000} has 302 digits. For very large values of n we replace the number 0.30103 by a more accurate value, namely 0.3010299957. For example, 2^{100000} has 30103 digits and not 30104.

To students familiar with the theory of logarithms, there is no mystery at all about this formula – the number 0.30103 is nothing but the logarithm of 2 to base 10. It is the number c for which

$$10^c = 2.$$

Solving the equation (we shall not go into the question of how this is done), we find that $c = 0.3010299957\ldots$.

⋆ ⋆ ⋆

The same investigation may be done with the powers of 3. Here, we solve the equation

$$10^c = 3.$$

The solution turns out to be $c = 0.4771212547\ldots$, or roughly 0.4771. We see therefore that if n is not too large, then

$$\#(3^n) = 1 + [0.4771 \times n].$$

Thus, for example, the number of digits in 3^{100} is $1 + [47.71]$ or 48, and the number of digits in 3^{1000} is $1 + [477.1]$ or 478. If $n > 1\,00\,000$, then we need to use the more accurate value given above.

10.3 First digit

In this section, we examine the powers of 2 and 3 from the reverse direction – from the "front" rather than from the "rear end". We consider only the decimal forms of 2^n and 3^n, and denote their first digits by b_n and c_n, respectively.

Example $b_5 = 3$, since $2^5 = 32$; $b_{10} = 1$, since $2^{10} = 1024$; $c_5 = 2$, since $3^5 = 243$; $c_{10} = 5$, since $3^{10} = 59049$; and so on.

If we prepare a table of values taken by b_n and c_n, we do find a pattern, but it is none too obvious. Considering only the first two hundred powers of 2 (from 2^1 till 2^{200}), we find that the frequency distribution of their first digits is as follows:

k	1	2	3	4	5	6	7	8	9
frequency [2]	60	36	24	20	16	13	11	11	9

(We have written "frequency [2]" as a reminder to show that the table refers to the powers of 2. More such notations, "frequency [3]", and so on, are used in the tables below.)

The first thing that strikes us is that the distribution is not uniform. Intuitively, we might have expected the digits 1, 2, 3, ..., 9 to occur equally often as the first digit, but this is clearly is not the case.

The same experiment may be done for the first digits of the first two hundred powers of 3. We obtain the following frequency table.

k	1	2	3	4	5	6	7	8	9
frequency [3]	59	36	25	19	16	13	11	11	10

The result is remarkable: *the first-digit distribution for the powers of 3 is practically the same as that for the powers of 2!* Why should this be so?

The question which ought to cross our minds next is whether the distribution of first digits for the powers of other integers, say 5 and 7, is similar to the above distributions. The answer is: *yes*.

k	1	2	3	4	5	6	7	8	9
frequency [5]	60	35	25	19	16	14	12	10	9
frequency [7]	62	33	26	20	15	13	12	10	9

This display of evidence should reconcile us to believing that, on the average, the first-digit distribution is the same for all power sequences. To progress towards understanding why this happens, we take the averages for the four distributions and obtain the following table, which gives the proportion, p_k, of powers for which the digit k occurs as the first digit ($k = 1, 2, 3, \ldots, 9$). The values have been rounded-off to 3 d.p.

k	1	2	3	4	5	6	7	8	9
p_k	.301	.175	.125	.098	.079	.067	.058	.053	.046

We now make a cumulative frequency table using these values of p_k. Let CF_k denote the cumulative frequency (CF_k is the sum of $p_1, p_2, \ldots, p_k$). The result is shown below (with values of CF_k rounded-off to 3 d.p.).

k	1	2	3	4	5	6	7	8	9
CF_k	.301	.476	.601	.699	.778	.844	.901	.954	1.0

For readers familiar with logarithms, the table should instantly strike a chord: CF_k is practically the same as $\log_{10}(k+1)$ (the "common logarithm" of $k+1$)! The values are not exactly the same, but we may expect closer agreement if we choose a larger number of powers than only the first two hundred. For comparison, we give below the values of $\log_{10} k$ for $k = 2, 3, 4, \ldots, 9$.

k	2	3	4	5	6	7	8	9	10
$\log_{10} k$	.301	.477	.602	.699	.778	.845	.903	.954	1.0

The closeness of the two sets of numbers is astonishing!

We leave the reader with the task of supplying an explanation for this apparent coincidence.

10.4 Coins and weights

Back to the bank! Or, mixing metaphors, to the market; we shall deal with weights rather than coins. Consider a shop where weights are kept only in the following measures: 1 kg, 2 kg, 4 kg, 8 kg, 16 kg, ... (all powers of 2). Can each integer weight be measured using these weights, if no measure may be used more than once? The answer is well known: *Every integer weight can be measured, and in precisely one way.* Otherwise stated:

> *For each positive integer N, there is precisely one solution to the equation $N = 2^a + 2^b + \cdots$ in distinct non-negative integers $a, b, \ldots$.*

Example We have $3 = 2^1 + 2^0$, $5 = 2^2 + 2^0$, $6 = 2^2 + 2^1$, $7 = 2^2 + 2^1 + 2^0$, $30 = 2^4 + 2^3 + 2^2 + 2^1, \ldots,$

The expression $N = 2^a + 2^b + \cdots$ gives the "base-2 expansion" of N. Thus, we have the following:

$$3 = 2^1 + 2^0 = (11)_2, \quad 5 = 2^2 + 2^0 = (101)_2, \quad \ldots.$$

Using this terminology, the above statement reads: *Every positive integer has a unique base-2 expansion.*

⋆ ⋆ ⋆

We move on to the powers of 3; the weight measures now are 1 kg, 3 kg, 9 kg, 27 kg, ... We clearly cannot cope if we have just one copy of each measure—we are unable even to measure 2 kg! Nor can we measure 5 kg, 6 kg or 7 kg. But a pretty result awaits us once we realize that weight measures may be used on both sides of the balance. Thus, 2 kg can be measured by placing the 1 kg measure along with the given object in one pan and the 3 kg measure in the other pan; 5 kg can be measured by placing the 1 kg and 3 kg measures with the given object on one side and the 9 kg measure on the other side; and so on. These schemes work because of the relations $2 = 3 - 1$ and $5 = 9 - 3 - 1$. Observe that 6 kg can be measured via $6 = 9 - 3$, and 7 kg via $7 = 9 - 3 + 1$. In general, we have the following statement:

> *For each positive integer N, there is precisely one solution to the equation $N = 3^a \pm 3^b \pm \cdots$ in distinct non-negative integers $a, b, \ldots$.*

Example We have $100 = 81 + 27 - 9 + 1$, $150 = 243 - 81 - 9 - 3$, $500 = 729 - 243 + 27 - 9 - 3 - 1$, and so on.

10.5 Repeated division

Here is a beautiful result obtained when we consider repeated division by 2.

For any number x, let $R(x)$ denote the integer *closest* to x; e.g., $R(2.7) = 3$, $R(3.1) = 3$. When the decimal part of x is 0.5, there are two integers which are 'closest'; we take $R(x)$ to be the greater one; e.g., $R(2.5) = 3$, $R(6.5) = 7$.

We select any positive integer n and find the following sum:

$$R\left(\frac{n}{2}\right) + R\left(\frac{n}{4}\right) + R\left(\frac{n}{8}\right) + R\left(\frac{n}{16}\right) + \cdots.$$

The sum may seem like an unending one, but this is not so; after a short while, all the terms become 0.

Example 1 Let $n = 25$; then $R(n/2) = 13$, $R(n/4) = 6$, $R(n/8) = 3$, $R(n/16) = 2$, $R(n/32) = 1$, $R(n/64) = 0$, $R(n/128) = 0, \ldots$; all succeeding terms are 0, and we have $13 + 6 + 3 + 2 + 1$, which equals 25.

Example 2 Let $n = 29$; then $R(n/2) = 15$, $R(n/4) = 7$, $R(n/8) = 4$, $R(n/16) = 2$, $R(n/32) = 1$, $R(n/64) = 0$, $R(n/128) = 0, \ldots$; all succeeding terms are 0, and we have $15 + 7 + 4 + 2 + 1$, which equals 29.

In each case, the sum turned out to be the original number. This is not a numerical coincidence! It turns out that we always have

$$n = R\left(\frac{n}{2}\right) + R\left(\frac{n}{4}\right) + R\left(\frac{n}{8}\right) + R\left(\frac{n}{16}\right) + R\left(\frac{n}{32}\right) + \cdots$$

The proof, while not difficult, involves a few subtleties, but we shall not elaborate further upon the matter.

★ ★ ★

If we round *down* rather than to the nearest integer, then we obtain another curious result. Recalling that the symbol $[x]$ means the greatest integer less than or equal to x, we compute the quantity $g(n)$ defined by

$$g(n) = \left[\frac{n}{2}\right] + \left[\frac{n}{4}\right] + \left[\frac{n}{8}\right] + \left[\frac{n}{16}\right] + \left[\frac{n}{32}\right] + \cdots.$$

Since we are always rounding down, and since earlier the sum had been n, we should expect $g(n)$ to be less than n; and this is so.

Example Consider the values $n = 20, 30$:

$$g(20) = 10 + 5 + 2 + 1 + 0 = 18 < 20,$$
$$g(30) = 15 + 7 + 3 + 1 + 0 = 26 < 30.$$

In both cases we have $g(n) < n$. This turns out to be true for all n. So the relation $g(n) \leq n - 1$ always holds good.

For which values of n, if any, is $g(n)$ equal to $n - 1$? The answer is unexpected and pleasing: $g(n) = n - 1$ precisely when n is a power of 2. For example, consider the values $n = 16$, 32 and 64:

$$g(16) = 8 + 4 + 2 + 1 + 0 = 15 = 16 - 1,$$

$$g(32) = 16 + 8 + 4 + 2 + 1 + 0 = 31 = 32 - 1,$$
$$g(64) = 32 + 16 + 8 + 4 + 2 + 1 + 0 = 63 = 64 - 1,$$

and so on. Similarly, $g(n) = n - 2$ *precisely when n is a sum of two unequal powers of 2.* As an example, consider the values $n = 20$, 36 and 48. Each of these is a sum of two unequal powers of 2 $(20 = 16 + 4,\ 36 = 32 + 4,\ 48 = 32 + 16)$. We already know that $g(20) = 18 = 20 - 2$; and now

$$g(36) = 18 + 9 + 4 + 2 + 1 + 0 = 34 = 36 - 2,$$
$$g(48) = 24 + 12 + 6 + 3 + 1 + 0 = 46 = 48 - 2.$$

In each instance we have $g(n) = n - 2$. It is not too hard to guess that $g(n) = n - 3$ precisely when n is a sum of three unequal powers of 2; and so on. We leave it to the reader to explore the "and so on" more fully.

★ ★ ★

The function $g(n) = [n/2] + [n/4] + \cdots$ turns up in another, very different context. Consider the product of all the positive integers less than or equal to n, namely:

$$1 \times 2 \times 3 \times \cdots \times (n - 1) \times n.$$

The product is denoted by $n!$ and pronounced as 'n factorial'. It arises frequently in problems of enumeration. Computation yields $1! = 1$, $2! = 2$, $3! = 6$, $4! = 24$ and $5! = 120$. For $n = 36$, we get an enormous number

$$36! = 371993326789901217467999448150835200000000.$$

The sequence of factorial numbers obviously increases with great speed; thus, the number 1000! has 2568 digits!

Let us now find, for each n, the exponent of the largest power of 2 that divides $n!$. For example, when $n = 5$ we have $5! = 120 = 2^3 \times 15$, and the exponent is 3; when $n = 7$ we have $7! = 5040 = 2^4 \times 315$, and the exponent is 4. In general, we have the following surprising result:

> *The exponent of the largest power of 2 which divides $n!$ is $g(n)$; that is, $2^{g(n)}$ divides $n!$, and no higher power of 2 divides $n!$.*

For example, for $n = 36$ we get

$$36! = 2^{34} \times 3^{17} \times 5^{8} \times 7^{5} \times 11^{3} \times 13^{2} \times 17^{2} \times 19 \times 23 \times 29 \times 31,$$

so the relevant exponent is 34. (Earlier we had found that $g(36) = 34$.)

Here is a pretty consequence of the above: *$n!$ is a multiple of 2^{n-1} if and only if n is a power of 2.* The reader is invited to fill in the details of the proof.

★ ★ ★

In the case of repeated division by 3 too, there are patterns to be discovered; but we leave this pleasant task of exploration to the reader. We mention only the following general result which holds for all primes p and all positive integers n: *The exponent of the largest power of p that divides $n!$ is given by the sum*

$$\left[\frac{n}{p}\right] + \left[\frac{n}{p^2}\right] + \left[\frac{n}{p^3}\right] + \cdots.$$

10.6 The Tower of Hanoi

The following puzzle is well known, but the unexpected occurrence of the powers of 2 in its analysis is most pleasing. (It is also known as the "Tower of Brahma".)

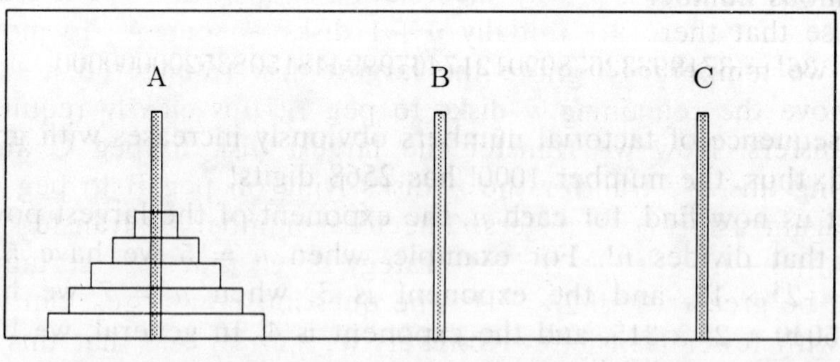

Figure B.10.2 *The tower of Hanoi (with five disks)*

The puzzle consists of three pegs, labelled A, B and C, and n disks where n is a given integer, with each disk having a hole in its centre. The disks are all of different sizes and initially they are all on peg A, in order of increasing size – the largest, disk #n, at the bottom, and the smallest, disk #1, at the top. The object of the puzzle is to move all the disks to peg C, using peg B as an intermediate junction, with the following constraint: *at no stage must any disk lie upon a smaller disk.*

Let h_n denote the least number of transfers needed to make the required change. Then, trivially, $h_1 = 1$ and $h_2 = 3$; for 2 disks may be moved thus – move disk #1 to B, then disk #2 to C, then disk #1 to C. Experimentation shows that $h_3 = 7$.

We find out soon enough that a very simple recurrence relation exists for the h-sequence; namely,

$$h_{n+1} = 2h_n + 1.$$

Using the formula, we get

$$h_3 = (2 \times 3) + 1 = 7, \quad h_4 = (2 \times 7) + 1 = 15,$$

and so on. Therefore, the sequence of h-values is as follows:

$$1,\ 3,\ 7,\ 15,\ 31,\ 63,\ 127,\ 255,\ \ldots.$$

Each number is 1 short of a power of 2! Using the notation developed in Part A, the h-sequence may be described as $\{2^n - 1\}$.

⋆ ⋆ ⋆

It is not hard to see why the relation $h_{n+1} = 2h_n + 1$ is true. Suppose that there are initially $n+1$ disks on peg A. To move them, we temporarily ignore the largest disk (numbered $n+1$) and move the remaining n disks to peg B; this clearly requires h_n transfers. Now we transfer the largest disk to peg C and, following this, we transfer the n disks lying on peg B to peg C, once again using h_n transfers. The total number of transfers is thus $h_n + 1 + h_n = 2h_n + 1$. This strategy shows that h_{n+1} certainly cannot be greater than $2h_n + 1$. The question now arises, can we do it with fewer transfers? However, we quickly see that this is not possible; for the largest disk cannot be shifted without first moving the n disks lying on top of it. This requires h_n transfers and, once the largest disk has been moved, a further h_n transfers are needed; and so on. So the recurrence relation is a genuine one.

Remark We note, in passing, that the 4-pegs version of the Tower of Hanoi is much more difficult to analyze than the 3-pegs version. No simple recurrence relation seems to exist for the least number of transfers needed in this case.

10.7 Mixing the powers of 2 and 3

Let the powers-of-2 and powers-of-3 sequences be mixed together and displayed as a single sequence, after rearranging the numbers in ascending order. We obtain the following sequence (note that we have left out the '1'):

$$\langle 2,\ 3,\ 4,\ 8,\ 9,\ 16,\ 27,\ 32,\ 64,\ 81,\ \ldots\rangle.$$

Two questions of interest which strike us are

(a) We see some pairs of consecutive integers occurring in the mixed sequence: $(2,3)$, $(3,4)$ and $(8,9)$. *Are there any more pairs of consecutive integers in the mixed sequence?*

(b) In the sequence 2, 3, 4, 6, 8, 9, 16, 27, 32, 64, 81, ..., we replace each power of 2 by the letter A, and each power of 3 by the letter B. The sequence now reads

$$A,\ B,\ A,\ A,\ B,\ A,\ B,\ A,\ A,\ B,\ \ldots.$$

The question now is: *Is there a simple rule by which we can determine whether the nth term is an A or a B?*

The solution to (b) involves the use of logarithms, and we shall not go into it here.

The answer to (a) is surprising: *there are no further such pairs of consecutive integers!* – the three pairs listed are the only such pairs. We sketch a proof of this claim below.

⋆ ⋆ ⋆

Suppose that m and n are positive integers such that 2^m and 3^n differ by 1. Then, there are two possibilities that need to be considered.

Case 1. $2^m - 3^n = 1$ The equation implies that $2^m - 1 = 3^n$. Now, the powers of 2 when divided by 3 leave remainders of 2 and 1 in alternation, and it is the *even* powers of 2 which leave a remainder of 1:

$$\text{Rem}(2^2 \div 3) = 1, \quad \text{Rem}(2^4 \div 3) = 1, \quad \ldots.$$

So, from the equation $2^m - 3^n = 1$, we infer that m is an even number. Write $m = 2k$, where k is a positive integer. Then, using the well known factorization for a difference of two squares ($x^2 - y^2 = (x-y)(x+y)$), we have

$$2^{2k} - 1 = 3^n, \quad \therefore \quad (2^k - 1)(2^k + 1) = 3^n.$$

So the product of $2^k - 1$ and $2^k + 1$ is a power of 3. Since 3 is a prime number, the only factors of a power of 3 are lesser powers of 3. This implies that both $2^k - 1$ and $2^k + 1$ are powers of 3. The two numbers differ by 2, so we ask: *which two powers of 3 differ by 2?* Clearly, the only such powers are 1 and 3. There can be no others, for the gap between the consecutive powers of 3 steadily increases as the powers grow larger. So we have $2^k - 1 = 1$ and $2^k + 1 = 3$, which means that $2^k = 2$ or $k = 1$, and so $m = 2k = 2$, $2^m = 4$, $3^n = 3$, $n = 1$. So $(3, 4)$ is the only pair of consecutive integers in the mixed sequence where the power of 3 precedes the power of 2.

Case 2. $2^m - 3^n = -1$ Now, we consider the case when the power of 2 precedes the power of 3. Write the equation as $2^m + 1 = 3^n$. Arguing as we did above, at the start, we infer that m is odd. (The equation implies that 2^m leaves a remainder of 2 when divided by 3, and this means that m is odd.) Next, we write the equation as $2^m = 3^n - 1$. We now distinguish between two subcases: n odd, and n even.

Suppose that n is odd. Then, the number $3^n - 1$ factorizes in an obvious manner:

$$3^n - 1 = (3-1)(3^{n-1} + 3^{n-2} + \cdots + 3^1 + 1).$$

Consider the bracketed term on the right, $(3^{n-1} + \cdots + 3 + 1)$; it contains n terms, and since n by supposition is odd, it is the sum of an odd number of odd numbers. Therefore, it is odd, which means that $3^n - 1$ is 2 times an odd number, and therefore also

$$2^m = 2 \times \text{an odd number}.$$

This shows that 2^m is not divisible by 4, so m is less than 2, therefore $m = 1$, $2^m + 1 = 3$, $n = 1$. The pair obtained under this subcase is thus $(2, 3)$.

Next we suppose that n is even, say $n = 2k$, where k is a positive integer. The approach now is similar to that used in Case 1: we factorize the number $3^{2k} - 1$ and write

$$2^m = (3^k - 1)(3^k + 1).$$

Since 2 is a prime number, the only factors of 2^m are lesser powers of 2, therefore both $3^k - 1$ and $3^k + 1$ are powers of 2; also, they differ by 2. The only two powers of 2 which differ by 2 are 2 and 4, so we have $3^k - 1 = 2$ and $3^k + 1 = 4$, or $3^k = 3$, $k = 1$, $m = 2k = 2$, $3^m = 9$, $2^n = 8$. We have now recovered the pair $(8, 9)$.

As all the relevant possibilities have been considered, it follows that $(2, 3)$, $(3, 4)$ and $(8, 9)$ are the only pairs of consecutive integers in the mixed powers-of-2-and-3 sequence.

⋆ ⋆ ⋆

More generally, we may ask for the solutions in integers $a, b, m, n > 1$ of the equation

$$a^m - b^n = 1.$$

We impose the condition $a, b, m, n > 1$ to avoid listing trivial solutions. [If for instance $n = 1$, then the equation reads $a^m = b + 1$, which obviously has infinitely many solutions (solutions like $2^{10} = 1023 + 1$) and these solutions clearly would not be of much interest to anyone!] We know that $a = 3$, $b = 2$, $m = 2$, $n = 3$ is a solution, and we have just shown that there are no others with a and b equal to 2 and 3 in some order. Are there other solutions? The reader may well be surprised to learn that this problem, known to mathematicians as *Catalan's problem*, is considered to be extremely difficult; indeed, it is currently unsolved. The numerical evidence is that there are no solutions other than the one given above, but we are still far from obtaining a proof of this statement.[1]

10.8 Sums of consecutive integers

Can every positive integer be written as a sum of two or more consecutive integers? Consider, for example, the integers 10, 20 and 30; we have

$10 = 1 + 2 + 3 + 4,$

[1]This problem was solved in September 2002 by Preda Mihăilescu, a mathematician from Romania. Catalan's conjecture is now a theorem.

$$20 = 2 + 3 + 4 + 5 + 6,$$
$$30 = 4 + 5 + 6 + 7 + 8.$$

Thus, the integers 10, 20, 30 can be so expressed. But some integers defy such a representation! Consider, for instance, the number 4. Since $1 + 2 < 4$ and $2 + 3 > 4$, no such expression is possible for 4. Also, trivially, 1 and 2 cannot be so expressed. The question of enumerating the integers, which are so representable, certainly seems an interesting one. An analysis reveals the following:

The positive integers which are not *representable as sums of two or more consecutive integers are precisely the powers of 2.*

A pretty and unexpected result indeed! And, fortunately, it is also fairly easy to prove—unlike many of the results we have had to quote in this and earlier chapters. The proof is given below.

Let n be a given positive integer, and suppose that n can be expressed as a sum of b consecutive integers, starting with a (here $a \geq 1$ and $b \geq 2$). Then, we have

$$n = a + (a+1) + (a+2) + \cdots + (a+b-2) + (a+b-1).$$

The sum on the right side can be rewritten using the standard trick—that of reversing the terms. We have,

$$n = (a+b-1) + (a+b-2) + (a+b-3) + \cdots + (a+1) + a.$$

Adding the two rows column by column, we get

$$2n = b(2a + b - 1).$$

This may be regarded as an equation in a and b, to be solved once n is given.

For instance, suppose that $n = 30$. Then $b(2a+b-1) = 60$, so b and $2a+b-1$ are numbers whose product is 60. Observe that their sum is $2a + 2b - 1$, an *odd* number. This implies that one of the numbers is odd and the other is even; also, $2a + b - 1 > b$. So we write 60 as a product of an odd number and an even number, set b equal to the smaller factor and then solve for a. For instance, we may write 60 as 5×12. This gives $b = 5$ and $2a + 5 - 1 = 12$ or $a = 4$; so 30 is the sum of 5 numbers starting with 4:

$$30 = 4 + 5 + 6 + 7 + 8.$$

We may also write 60 as 3×20; this gives $b = 3$, $2a + 3 - 1 = 20$, $\therefore\ a = 9$, and

$$30 = 9 + 10 + 11.$$

So the formula $2n = b(2a + b - 1)$ allows us to find all possible representations, by factorizing $2n$ in all possible ways as an odd number times an even number.

In the general case, we argue as follows. Since $2n$ is equal to $b(2a + b - 1)$, and the sum of $b, 2a + b - 1$ is

$$b + (2a + b - 1) = 2(a + b) - 1,$$

which is *odd*, one of the numbers b, $2a + b - 1$ is odd and the other is even. Since b is greater than 1, and $2a + b - 1$ is greater than b and therefore greater than 1, $2n$ certainly contains an odd divisor which exceeds 1. This shows that $2n$ cannot be a power of 2, because the *only* odd divisor of a power of 2 is 1. So, no power of 2 is representable as a sum of two or more consecutive integers.

To complete the analysis we must show that if n is not a power of 2, then such a representation *is* possible. But this is easy: if n is greater than 1 and not a power of 2, then it has an odd factor greater than 1. Let b be the least such factor. Now a can be found from the equation

$$2a + b - 1 = \frac{2n}{b},$$

and the desired representation is at hand.

Exercises

10.8.1 Prove that

$$1 + 3 + 3^2 + 3^3 + \cdots + 3^n = \frac{3^{n+1} - 1}{2}$$

for all positive integers n.

10.8.2 Prove that

$$1 + 4 + 4^2 + 4^3 + \cdots + 4^n = \frac{4^{n+1} - 1}{3}$$

for all positive integers n.

10.8.3 Guess a formula for the sum

$$1 + a + a^2 + a^3 + \cdots + a^n$$

where a and n are arbitrary positive integers.

10.8.4 Investigate the sequence of remainders obtained when the powers of 2 are divided by 3. Show that the observed pattern is genuine.

10.8.5 Investigate the sequence of remainders obtained when the powers of 2 are divided by 5.

10.8.6 Investigate the sequence of remainders obtained when the powers of 2 are divided by 7.
Using your findings, show that there are no positive integers n for which $2^n + 1$ is divisible by 7.

10.8.7 Show that in the mixed powers-of-2-and-5 sequence, the only pair of consecutive integers is $(4, 5)$.

10.8.8 Show that in the mixed powers-of-2-and-7 sequence, the only pair of consecutive integers is $(7, 8)$.

10.8.9 Express 1000 and 1500 as sums of two or more consecutive integers.

10.8.10 Which positive integers can be written as sums of two or more consecutive odd numbers?

Example: The integers 8 and 21 can be so expressed ($8 = 3 + 5$ and $21 = 5 + 7 + 9$), but not 6 or 10.

Investigate the general problem.

Chapter 11

The Sequence of Primes

In this chapter, we study the prime numbers[1]

$$2,\ 3,\ 5,\ 7,\ 11,\ 13,\ 17,\ 19,\ 23,\ \ldots,$$

which have fascinated mathematicians since ancient times and continue to do so! Problems concerning the primes still excite and baffle today's mathematicians. This chapter describes some of these problems.

11.1 Infinitude of the primes

We start with the most basic question: *How long does the sequence of primes continue?* Long ago, the Greeks showed by a simple and yet very effective argument that the number of primes is infinite. There is nothing "obvious" about this fact, and what may seem to be the most natural approach towards showing the infinitude of primes has not succeeded till now.

The proof, as described by Euclid in *The Elements*, is based on self-contradiction. We have used this idea in an earlier chapter, to show the impossibility of finding two squares in the ratio 1 : 2. Here, we *assume* the existence of a largest prime number P, then show the absurdity of this assumption. Let N be the number given by:

$$N = (\text{product of all the primes till } P) + 1.$$

So if someone were to claim that 11 is the largest prime, then we would have $N = (2 \times 3 \times 5 \times 7 \times 11) + 1$, or $N = 2311$.

[1] Observe that we have not listed 1 as a prime number. The reason behind this is described later in this chapter.

Because of the "+1" in this definition, it is clear that N is not divisible by any of the primes till P; indeed, it leaves a remainder of 1 when divided by each of them.

What kind of number is N, then? Either it is prime or it is composite. It cannot be prime, because N exceeds P, and we have assumed that P is the largest prime. So it must be composite. But then what are its prime factors?—it cannot have any, because it is indivisible by all the available primes! This is an absurdity, but it follows inevitably from the initial assumption. The only way out is to accept that there are primes beyond P. So there can be no largest prime number, which is the same as saying that the number of primes is infinite. *Thus, the sequence of primes carries on for ever.*

Number play may help us to follow the argument better. If someone were to claim that 7 is the largest prime, then we would compute the number $N = (2 \times 3 \times 5 \times 7) + 1 = 211$, and ask, what kind of number is 211? It is not divisible by any of the primes 2, 3, 5 or 7, so either it is itself prime or it has prime factors which are not in the list 2, 3, 5, 7. In fact, it *is* prime. So 7 cannot be the largest prime number.

If we had started with $P = 13$, then we would compute

$$N = (2 \times 3 \times 5 \times 7 \times 11 \times 13) + 1 = 30031,$$

and this number is *not* prime: it is divisible by the two primes 59, 509 (indeed, $30031 = 59 \times 509$). So, we have two primes which are both larger than the supposed largest prime, 13.

This is Euclid's proof—a remarkable achievement! What may not be obvious at first sight is the way in which it does *exactly* what it sets out to do, with no wasted effort. Now we can explain the comment made above, that "... the most natural approach towards showing the infinitude of primes has not succeeded till now." The most natural approach would perhaps be to find a formula for the nth prime; or a formula for the next prime following a given one. Either approach would prove the infinitude of the primes—provided it is successful! Unfortunately, in the 2000 years since Euclid wrote *The Elements*, little progress has been made with either approach. Clearly, the two problems are difficult. We see now how simple and economical is the Greek approach, how little "waste" it has.

⋆ ⋆ ⋆

Another, rather elegant way in which the infinitude of the primes may be shown is the following. Let the numbers $B_1, B_2, B_3, \ldots$ be defined recursively as follows: $B_1 = 1$, and

$$\begin{aligned} B_2 &= B_1 + 1 = 2, \\ B_3 &= B_1 B_2 + 1 = 3, \\ B_4 &= B_1 B_2 B_3 + 1 = 7, \\ B_5 &= B_1 B_2 B_3 B_4 + 1 = 43, \end{aligned}$$

and so on; in general, B_n is 1 more than the product of all the preceding B_is. Observe that no two of the B_is share a common factor; for if $m > n$, then

$$B_m = (\text{some integer} \times B_n) + 1,$$

so a factor that divides both B_m and B_n would have to be a divisor of 1 as well. It follows from this that the common factor is 1.

Now observe that each B_i has associated with it, its own set of prime factors. The fact that the B-numbers have no factors in common shows that these sets are disjoint from one another. Since there are infinitely many B-numbers, it follows that there are infinitely many prime numbers.

⋆ ⋆ ⋆

The primes may be called the "building blocks" of the universe of numbers. An important fact about primes is the "fundamental theorem of arithmetic", according to which *the prime factorization of each integer is unique*, with variations possible at most in the order in which the primes are listed. For example, $30031 = 59 \times 509$, and this is the only way 30031 can be factorized into primes. There is no variation possible in the exponents either; thus $72 = 2^3 \times 3^2$, and there is no other way of writing 72 in the required form.

Now we see why 1 is not listed as a prime. If we had declared 1 to be prime, then 30031 could be written as a product of primes in infinitely many ways as follows:

$$\begin{aligned} 30031 &= 1 \times 59 \times 509, \\ 30031 &= 1^2 \times 59 \times 509, \\ 30031 &= 1^3 \times 59 \times 509, \end{aligned}$$

and so on. Each way would have to be regarded as different, and we would lose the fundamental theorem of arithmetic. This is not acceptable! The fundamental theorem has too many important uses for us to lose it in so trivial a manner, so we take the easy way out and declare that 1 is not a prime number. (Of course it is not composite either; rather, it is a *unit*. In the set of integers, the units are ± 1; these are the only integers whose reciprocals too are integers.)

★ ★ ★

Since the primes are infinite in number, so are the composite numbers. Indeed, each of the prime-power sequences

$$2,\ 4,\ 8,\ 16,\ 32,\ 64,\ 128,\ \ldots,$$
$$3,\ 9,\ 27,\ 81,\ 243,\ 729,\ \ldots,$$
$$5,\ 25,\ 125,\ 625,\ 3125,\ \ldots$$

contains infinitely many composite numbers; and there are lots more of them! It is not hard to check that the composite numbers are more numerous than the primes. Thus, among the integers till 10,000, only 1229 are prime – less than $12\frac{1}{2}\%$. Table B.11.1 gives some data on the distribution of primes. The evidence that the primes gradually thin out is plainly visible. Yet, considered as wholes, the sets of prime numbers and composite numbers are both infinite.

Table B.11.1 *Distribution of the primes*

N	# primes $\leq N$	approx. density
10	4	0.400
100	25	0.250
1000	168	0.168
10000	1229	0.123
100000	9592	0.096
1000000	78498	0.078
10000000	664579	0.066
100000000	5761455	0.058

★ ★ ★

The very first prime, 2, is even, and it is the only even prime (the even numbers after 2 are clearly not prime), so 2 is the oddest prime of all! The only two consecutive numbers which are both prime are 2 and 3. In contrast, lists of consecutive composite numbers of any desired length may be constructed. Thus, the list 122, 123, 124, 125 has four consecutive composite numbers, and the list 722, 723, 724, 725, 726 has five consecutive composite numbers. Indefinitely large lists of this kind can be constructed as follows. Let n be any integer greater than 3, and let N be the number $1 \times 2 \times 3 \times \cdots \times n$ (this is generally written as $n!$). Now, consider the following list of $n - 1$ consecutive numbers:

$$N + 2,\ N + 3,\ N + 4,\ N + 5,\ \ldots,\ N + n - 1,\ N + n.$$

Each of these is composite: $N + 2$ has 2 as a factor, $N + 3$ has 3 as a factor, ..., $N + n$ has n as a factor. So we have a list of $n - 1$ consecutive numbers, which are all composite. And, thereby, a means of constructing indefinitely long lists of consecutive composite numbers. Another way of expressing this is: *The gap between consecutive primes can be indefinitely large.*

The above result starkly illustrates the extreme irregularity of the primes. Yet, looked at "in the large", the primes possess a strange regularity, and this may be shown as follows. Let $P(n)$ denote the 2^n-th prime; e.g., $P(1) = 3$ is the 2nd prime, $P(2) = 7$ is the 4th prime, $P(10) = 8161$ is the 1024th prime, and so on. If we plot a graph of $\log_{10} P(n)$ versus n, leaving the points as dots and not joining them up, here is what we get. The near straightness of the line on which the dots lie, is striking.

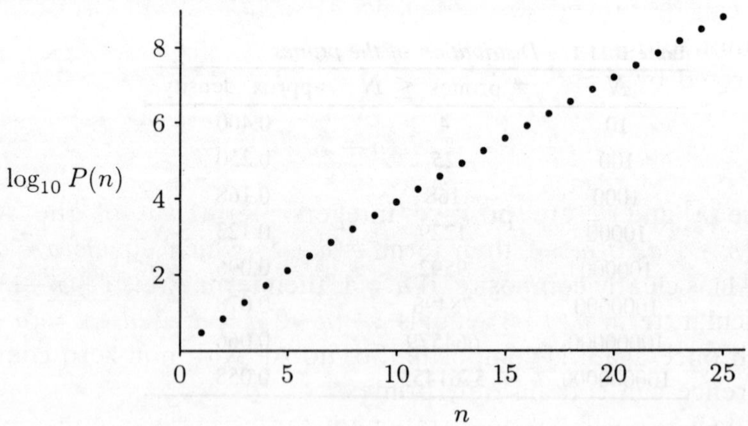

Figure B.11.1 *Graph of* $\log_{10} P(n)$ *versus* n

⋆ ⋆ ⋆

The fact there are infinitely many primes prompts us to ask

for a formula that will generate all of them; or, moderating our ambitions, a formula that will yield infinitely many primes. But this quest proves to be a frustrating one!—no worthwhile formula has ever been found that will do these tasks. The nicest formula discovered till now is Euler's: $x^2 + x + 41$, which is prime for no less than eighty consecutive values of x (namely, $x = -40, -39, \ldots, 0, 1, \ldots, 38, 39$); each prime value is taken twice. But it is composite for $x = 40$.

Other formulas of this kind, but less impressive, are the following:

- $x^2 + x + 11$ (prime for 20 consecutive values of x, namely of $x = -10, -9, -8, \ldots, 8, 9$);
- $x^2 + x + 17$ (prime for 32 consecutive values of x, namely $x = -16, -15, -14, \ldots, 14, 15$).

Observe that the three formulas listed above are all of the form $x^2 + x + k$, for some integer k. It turns out that the best possible quadratic function of this form is given by Euler's formula, $x^2 + x + 41$; but the proof of this statement is very difficult.

⋆ ⋆ ⋆

It can be proved easily that no AP contains *only* primes if the common difference d is non-zero. To see why, consider the AP generated by a and d,

$$a,\ a+d,\ a+2d,\ a+3d,\ a+4d,\ \ldots,$$

where a and d are positive integers. Term $\#n$ of the AP is $a+(n-1)d$. If $a > 1$ then term $\#(a+1)$ which equals $a+ad$ or $a(d+1)$ is clearly composite. If $a = 1$, then term $\#n$ is $1+(n-1)d$. In particular, term $\#(d+3)$ equals $1+(d+2)d = d^2+2d+1 = (d+1)^2$, which once again is composite. So, no AP with non-zero common difference can contain only primes.

In fact, a much stronger statement can be made: a non-constant polynomial with integer coefficients cannot generate *only* prime values. This had been stated, without proof, in Chapter 6; an outline of the proof is given later in this chapter. Note that the statement proved above is a particular case of this statement (for the AP $a,\ a+d,\ a+2d,\ a+3d, \ldots$ is generated by the linear polynomial $dx + a$).

A non-constant AP cannot consist only of primes, but it can certainly contain infinitely many prime numbers. The most impressive result in this respect is the following, due to Peter Gustav Lejeune Dirichlet:

> *If a and d are coprime integers, then the* AP a, $a+d$, $a+2d$, $a+3d$, ..., *contains infinitely many primes.*

So we are assured, for instance, that the AP 3, 7, 11, 15, 19, ..., contains infinitely many primes. Here $a = 3$, $d = 4$, and these two integers are coprime. We are also assured that there are infinitely many primes whose units digit is 1 (think of the AP 1, 11, 21, 31, ...,); infinitely many primes whose units digit is 3; infinitely many primes whose last two digits are 01 (think of the AP 1, 101, 201, 301, ...,); infinitely many primes whose last five digits are 00003 (think of the AP 3, 100003, 200003, 300003, ...,); and so on. Clearly, there is a lot packed into Dirichlet's statement!

However, Dirichlet's theorem is *very* hard to prove. For particular APs simpler proofs may be possible. For instance, consider the AP 3, 7, 11, 15, ..., generated by the formula $4k-1$. To show that this AP contains infinitely many primes, we argue as the Greeks did. Suppose that amongst the numbers 3, 7, 11, 15, ..., there is a largest prime number, say P. Now, consider the number N given by

$$N = 4(3 \times 7 \times 11 \times 15 \times \cdots \times P) - 1.$$

This number is of the form $4k - 1$ and therefore belongs to the AP. Also, it is not divisible by any number in the AP till P. So either it is prime, or it is composite with prime factors larger than P. In the former case, we obtain a prime number in the AP which exceeds P and this is contrary to supposition. In the latter case, we ask: *what is the nature of its prime factors? Can the prime factors all be of the form* $4k+1$*?* This is not possible, for if all the prime factors of N were of this form, then N itself would be of this form. (To see why, observe that the product of the numbers $4n+1$ and $4n'+1$ is

$$16nn' + 4n + 4n' + 1 = 4(4nn' + n + n') + 1,$$

which is of the type $4k+1$.) So the prime factors of N cannot all be of the type $4k+1$; at least one prime factor must belong to the given AP. This factor cannot be less than P, because, as already observed, N is indivisible by numbers in the AP less than P. The only way out now is to accept that the AP contains prime numbers beyond P. Therefore, the AP contains infinitely many primes.

⋆ ⋆ ⋆

No AP with a non-zero common difference can contain only primes. This statement may prompt us to look for *finite* lists of primes in arithmetic progression. For instance, there are prime triples—lists such as 3, 13, 23. Are there infinitely many such triples? The answer is not known! We may look for longer lists. An example of a four-term list is 13, 43, 73, 103, and here are two lists containing six terms:

a) 7, 37, 67, 97, 127, 157;

b) 107, 137, 167, 197, 227, 257.

Many such lists are known, some of impressive length; indeed, "world records" exist for such lists! Is there an upper limit to the length of such a list? The answer to this question too is not known. Indeed, such questions multiply with bewildering ease, and one is quickly overwhelmed by their number. In the opinion of many mathematicians, among them Paul Erdős, mathematics is not yet ready to answer such questions.

A few questions of this genre however *can* be answered. Consider the triplet 3, 5, 7: a three-term "prime AP" with the common difference 2. Is there another such triplet? No, and this is easy to prove. Clearly, any such triplet must contain only odd numbers. Now of any three consecutive odd numbers, one must be a multiple of 3. If this number is prime, then it must be 3 itself, so the remaining numbers are 5 and 7 (since 1 is disallowed, not being considered as a prime). So the only such triplet is 3, 5, 7.

⋆ ⋆ ⋆

We now sketch the proof of the assertion made earlier, that a non-constant polynomial with integer coefficients cannot generate only prime values. We shall give the proof only for quadratic functions. However, the general idea works for any such polynomial; we leave the details to the reader.

Let $f(x)$ be a quadratic function, say $f(x) = ax^2 + bx + c$ where a, b, c are integers with $a > 0$. (If $a < 0$, we consider the function $-f(x)$. If $a = 0$, then f is linear; we have already considered this case.) We shall show that $f(x)$ cannot take only prime values. By "completing the square" we get

$$f(x) = a\left(x + \frac{b}{2a}\right)^2 + \left(c - \frac{b^2}{4a}\right).$$

Since a squared quantity is never negative, we see that $f(x)$ achieves its lowest value when $x = -b/2a$, and steadily rises ("to infinity") when x goes beyond this. So it will happen that $f(x+1) > f(x) > 1$ for all x beyond some stage, say for all $x \geq d$ where d is some positive integer. Let $f(d) = k$; then $k = ad^2 + bd + c$ and $k > 1$. Consider the value $f(d+k)$. Since $d + k > d$, we have $f(d+k) > f(d) > 1$. Next,

$$\begin{aligned} f(d+k) &= a(d+k)^2 + b(d+k) + c \\ &= (ad^2 + bd + c) + (2adk + ak^2 + bk) \\ &= k + 2adk + ak^2 + bk = k(1 + 2ad + ak + b). \end{aligned}$$

So $f(d+k)$ is a multiple of k, and it exceeds k; therefore, $f(d+k)$ is composite. It follows similarly that the numbers

$$f(d+2k),\ f(d+3k),\ f(d+4k),\ \ldots$$

are all composite. So f generates infinitely many composite numbers. It also follows that f cannot generate only prime values; indeed that it cannot even generate only prime values from some point onwards. Exactly the same idea works for polynomials with degree greater than 2.

⋆ ⋆ ⋆

A curious family of primes is found in the set of *repunit numbers*. These are the numbers which, in base-10, contain only 1s. Our interest is in *prime repunits*, e.g., the prime 11. Let R_n denote the nth repunit:

$$R_n = \underbrace{111\ldots111}_{n \text{ ones}}.$$

It is easy to show that if n is composite, then so is R_n. So if we want prime repunits, then we need consider only the cases where n is prime. A computer search reveals that for n less than 100, the only values for which R_n is prime are

$$n = 2, \quad 19, \quad 23.$$

Factorizations of some non-prime repunits may be of interest—$R_3 = 3 \times 37$, $R_5 = 41 \times 271$, $R_7 = 239 \times 4649$, and for $n = 11$, 13 and 17:

$$\begin{aligned} R_{11} &= 21649 \times 513239, \\ R_{13} &= 53 \times 79 \times 265371653, \\ R_{17} &= 2071723 \times 5363222357. \end{aligned}$$

A rather non-obvious result about repunits is the following: *If p is a prime number greater than 5, then p is a divisor of R_{p-1}.*

Example 1 Let $p = 7$; then $R_6 = 111111 = 7 \times 15873$.

Example 2 Let $p = 13$; then $R_{12} = 13 \times 8547008547$.

The proof of this requires the famous and very important "little theorem" of Fermat's: *If p is a prime number and a is an integer not divisible by p, then $a^{p-1} - 1$ is a multiple of p.* We leave the study of the theorem to the reader.

★ ★ ★

A curious connection exists between the primes and the squares. To see the connection, we compute the the sequence of partial sums of the primes; that is, the numbers 2, $2+3 = 5$, $2+3+5 = 10$, $2+3+5+7 = 17$, $2+3+5+7+11 = 28$ and so on, obtained by cumulatively adding the primes. We obtain the following sequence of numbers:

$$\begin{array}{l} 2,\ 5,\ 10,\ 17,\ 28,\ 41,\ 58,\ 77,\ 100,\ 129,\ 160,\ 197, \\ 238,\ 281,\ 328,\ 381,\ 440,\ 501,\ 568,\ 639,\ 712,\ 791, \\ 874,\ 963,\ 1060,\ 1161,\ 1264,\ 1371,\ 1480,\ 1593, \\ 1720,\ 1851,\ \ldots. \end{array}$$

Observe that *between every two consecutive partial sums there lies a square*. Thus, between 2 and 5 lies 4; between 5 and 10 lies 9; between 10 and 17 lies 16; between 17 and 28 lies 25; between 28 and 41 lies 36; and so on. Will this always be the case? The answer is yes, and we can show it to be true in the following manner.

Let p_n denote the nth prime. The crucial fact about the primes that we use is that after the very first prime, 2, all the primes are odd, so the gap between consecutive primes after the first two is at least 2. This implies that $p_n \geq 2n-1$ for all n (e.g., $p_1 > 1$, $p_2 = 3$, $p_3 = 5$, $p_4 = 7$, $p_5 = 11 > 9$, $p_6 = 13 > 11$), with strict inequality for several n (e.g., $n = 1, 5, 6$). Since $1+3+5+\cdots+(2n-3)+(2n-1) = n^2$, we deduce that

$$p_1 + p_2 + p_3 + \cdots + p_n > n^2 \quad \text{for all } n.$$

We also deduce the following: if for any n we have $p_n < 2k+1$ for some k, then we also have $p_{n-1} < 2k-1$.

Let $s_n = p_1 + p_2 + \cdots + p_n$ denote the nth partial sum of the sequence of primes. We must show that between any two consecutive values of s_n, there exists a square. Suppose that the largest square less than or equal to s_n is k^2; then

$$k^2 \leq s_n < (k+1)^2.$$

We know that $s_n > n^2$, so n^2 is one of the squares less than or equal to s_n. Since k^2 is the largest such square, we get $n^2 \leq k^2$, or $n \leq k$.

Now suppose that s_{n+1} also lies in the interval between k^2 and $(k+1)^2$; then we have,

$$n^2 \leq k^2 \leq s_n < s_{n+1} = s_n + p_{n+1} \leq (k+1)^2.$$

Since $(k+1)^2 - k^2 = 2k+1$, these relations imply that $p_{n+1} < 2k+1$. This yields, in succession, $p_n < 2k-1$, $\therefore p_{n-1} < 2k-3$, $\therefore p_{n-2} < 2k-5, \ldots, p_2 < 2k-(2n-5)$ and finally $p_1 < 2k-(2n-3)$. (Note that since $k \geq n$ the last inequality is not an absurd one.) Therefore,

$$\begin{aligned} s_n &= p_n + p_{n-1} + p_{n-2} + \cdots + p_1 \\ &< (2k-1) + (2k-3) + \cdots + (2k-(2n-3)) \\ &< (2k-1) + (2k-3) + \cdots + 3 + 1 \;=\; k^2. \end{aligned}$$

So $s_n < k^2$, contradicting the initial supposition that k^2 is the largest square less than or equal to s_n. It is not possible that s_{n+1} too lies between k^2 and $(k+1)^2$. Therefore, $s_{n+1} > (k+1)^2$, and we see that the square $(k+1)^2$ lies between s_n and s_{n+1}.

The *only* fact about the primes used in the above proof is $p_{n+1} - p_n \geq 2$ for $n > 1$. Thus, the result proved above will hold for any sequence of positive integers for which the gap between consecutive members is at least 2.

★ ★ ★

The primes are infinite in number, but this does not mean that we know infinitely many of them! Large prime numbers have always been objects of great interest—perhaps because everyone loves large numbers! The search for large primes is usually restricted to particular families of numbers. We study two such families in the next section.

11.2 Mersenne and Fermat primes

Of the many families of primes, two stand out: primes of the form $2^n - 1$, and primes of the form $2^n + 1$. Historically, these have played an important role in the development of number theory.

Consider the numbers of the form $2^n - 1$; they are known as the "Mersenne numbers", after Father Marin Mersenne who stated in 1644 that $2^p - 1$ is prime when $p = 2, 3, 5, 7, 13, 17, 19, 31, 67, 127$ and 257, but composite for all other prime values of p less than 257 (there are 44 such primes). Mersenne's claim is an impressive one, considering the size of the numbers involved ($2^{257} - 1$ is a 78-digit number). Let us see how far Mersenne's claim may be verified.

We write M_p for $2^p - 1$. The values taken by M_p when $p = 2, 3, 5$ and 7 are 3, 7, 31 and 127 respectively, and these are all prime; next, $M_{11} = 2047 = 23 \times 89$ is not prime; $M_{13} = 8191$ is prime, and so is $M_{17} = 131071$. So far, so good! However, Mersenne's claim is faulty: M_p is prime for $p = 61, 89$ and 107 (these numbers do not feature on his list), and composite for $p = 67$ and 257 (both these feature on the list). Despite these errors, Mersenne's feat in producing the list must be considered extremely remarkable. It is not quite clear how he prepared the list.

It is easy to see that if n is composite, then so is M_n. This is due to the fact that if x and n are integers greater than 1, then $x^n - 1$ is divisible by $x - 1$:

$$x^n - 1 = (x - 1)(x^{n-1} + x^{n-2} + \cdots + x + 1).$$

So if n is composite, say $n = mk$ where m and k exceed 1, then $x^n - 1$ is divisible by $x^k - 1$. In particular, M_{mk} is divisible by M_k and so is composite. Therefore, in studying the Mersenne numbers, we may restrict ourselves to examining only the cases where the exponent is prime.

⋆ ⋆ ⋆

Primes of the form $2^p - 1$ are called *Mersenne primes*. They can certainly astound us by their size! Thus, we have the following enormous prime number:

$$2^{127} - 1 = 170141183460469231731687303715884105727,$$

a number with 39 digits. (But, large as it is, there are infinitely many primes beyond it . . . .)

Are there infinitely many Mersenne primes? The answer to this question is not known. The list of Mersenne primes keeps growing, and every few years an addition is made to the list. Here is a list of primes p, for which M_p is prime, updated till 1997. The list is complete till the largest p shown.

2,	17,	107,	2203,	9689,	23209,	216091,
3,	19,	127,	2281,	9941,	44497,	756839.
5,	31,	521,	3217,	11213,	86243,	
7,	61,	607,	4253,	19937,	110503,	
13,	89,	1279,	4423,	21701,	132049,	

Two interesting results concerning the Mersenne primes may be stated here. The first concerns *perfect numbers*, numbers for which the sum of the proper divisors equals the number itself. Two examples of such numbers are 6 and 28. Note that

- the proper divisors of 6 are 1, 2 and 3, and their sum is 6;
- the proper divisors of 28 are 1, 2, 4, 7 and 14, and their sum is 28.

Now for the connection: *If P is a Mersenne prime, then the number $P(P+1)/2$ is perfect.* (The Greeks were aware of this fact.) So, from the Mersenne primes 3, 7, 31, 127, we obtain the following four perfect numbers:

$$\frac{3 \times 4}{2} = 6, \qquad \frac{31 \times 32}{2} = 496,$$
$$\frac{7 \times 8}{2} = 28, \qquad \frac{127 \times 128}{2} = 8128.$$

However, more can be said: *Every even perfect number is of the form $P(P+1)/2$ for some Mersenne prime, P.* This was first shown by Euler. (It was not known to the Greeks.)

Note that we have used the word *even*. Are there any odd perfect numbers? Amazingly, the answer to this question too is not known.

The other result is more of a curiosity: *The sum of all the divisors of a number, including the number itself, is a power of 2 if and only if it is of the form $P_1 P_2 \cdots P_k$ where P_1, P_2, ..., P_k are distinct Mersenne primes.*

Example Consider the number $93 = 3 \times 31$; its divisors are 1, 3, 31, 93, and their sum is 128, a power of 2. Or consider the number $889 = 7 \times 127$; its divisors are 1, 7, 127 and 889, and their sum is 1024, again a power of 2.

⋆ ⋆ ⋆

We now move on to a study of primes of the form $2^n + 1$. The following result is easy to establish: *If n has an odd divisor greater than 1, say k, then 2^n+1 is divisible by $2^{n/k}+1$.* For example, consider $n = 10$, which has the odd divisor 5; observe that $2^{10} + 1 = 1025$ is divisible by $2^2 + 1$ or 5. Or consider $n = 15$; this has the odd divisors 3 and 5, and we find that the number $2^{15} + 1 = 32769$ is divisible by $2^3 + 1$ or 9, as well as by $2^5 + 1$ or 33. More generally, we have the following identity, which is easy to check simply by multiplication: if n is odd, then

$$\frac{x^n + 1}{x + 1} = x^{n-1} - x^{n-2} + x^{n-3} - x^{n-4} + \cdots - x + 1.$$

The signs are alternately plus and minus. (No such identity holds when n is even; in this case, $x + 1$ does not divide $x^n + 1$.) In particular, if $n = mk$ and k is odd, then we have

$$\frac{2^n + 1}{2^m + 1} = 2^{n-m} - 2^{n-2m} + 2^{n-3m} - \cdots + 1.$$

Example Let $n = 12$, $k = 3$, $m = 4$, then

$$\frac{2^{12} + 1}{2^4 + 1} = 2^8 - 2^4 + 1.$$

So if $2^n + 1$ is to be prime, then n must have no odd factor greater than 1; i.e., n must be a power of 2. Primes of the form $2^n + 1$ are known as *Fermat primes*, named after the "greatest of the amateurs", Pierre Fermat.[2]

---

[2] Fermat was, by profession, a jurist for the Government of France. He made many beautiful discoveries in Number Theory, Analytic Geometry and Probability Theory.

The interest in such primes may be traced to a claim made by Fermat, that the numbers F_n defined by

$$F_n = 2^{2^n} + 1$$

are prime for $n = 0, 1, 2, 3, 4$, "and so on". The claim is certainly true for the values of n shown:

$$F_0 = 3, \quad F_1 = 5, \quad F_2 = 17, \quad F_3 = 257, \quad F_4 = 65537.$$

But the very next Fermat number is not prime, and this was first shown by Euler thus:

$$F_5 = 2^{32} + 1 = 4294967297 = 641 \times 6700417.$$

A curious (and amusing) aspect of Fermat's claim is that not even one prime has been found among the Fermat numbers after F_4! (It is conjectured that F_n is composite for all $n > 4$.)

Among the many fascinating properties of the Fermat primes is their connection with constructions in geometry using a ruler and a compass, first discovered by Karl Gauss in the 1790s (he was then a teenager). Apparently, it was this discovery that persuaded Gauss to pursue a career in mathematics (till then he was undecided)!

We first state what is meant by "constructions using a ruler and a compass". A ruler is simply an unmarked straight edge, and everyone knows what a compass is. The lack of markings on the ruler means that lengths can be transferred only via use of the compass. Using these two 'Euclidean instruments', we may: (a) draw a line joining two previously constructed points and (b) draw a circle with a previously constructed point as centre, passing through another previously constructed point. It is clear that any construction in Euclidean geometry consists of a repetition of these two actions.

The ancient Greeks were fascinated by constructions using a ruler and a compass, and they asked, *Is there a way, for an arbitrary integer $n \geq 3$, of constructing a regular n-sided polygon ("n-gon")?* The cases $n = 3$ (equilateral triangle), $n = 4$ (square), $n = 6$ (regular hexagon) and $n = 8$ (regular octagon) are familiar, and the constructions are straightforward. The case $n = 5$ (pentagon) is less familiar, but there is a pretty method here too (it was known to the Greeks). The question now is: *For which n can a regular n-gon be constructed using a ruler and a compass?* Note that if a construction is possible for some n, then it is also possible for

$2n$, simply via angle bisection. So we need to consider only odd values of n. It was Gauss who first discovered the following very surprising fact:

> *If P is a Fermat prime, then a regular P-gon may be constructed using a ruler and a compass.*

Thus, regular 5-gons and 17-gons are constructible using these two instruments.

More generally, it is true that *a regular n-gon may be constructed using a ruler and a compass if and only if n is of the form $2^k P_1 \cdots P_r$, where the P_i are distinct Fermat primes and k is a non-negative integer.*

Here, the word "distinct" is used in the sense that the prime factors must not be repeated. Thus, a regular polygon with 3 sides may be constructed, or one with $15 = 3 \times 5$ sides, or one with $2 \times 3 \times 17 = 102$ sides, or even one with $3 \times 257 = 771$ sides (!), but not one with $3^2 = 9$ sides, nor one with $2 \times 5^2 = 50$ sides.

This result settles an old problem of Euclidean geometry—that of constructing a regular 7-gon. Since 7 is prime but not a Fermat prime, the required construction is not possible.

⋆ ⋆ ⋆

Earlier, we had stated that the sum of the divisors of a number n is a power of 2 if and only if n is a product of distinct Mersenne primes. We find that an analogous result holds for Fermat primes too. It concerns the Euler phi-function $\phi(n)$ (also known as the "totient function"). We shall first define ϕ.

• **Euler's phi-function** Given a positive integer n, we define $\phi(n)$ to be the number of positive integers less than or equal to n and coprime to n.

Example We have $\phi(1) = 1$, $\phi(2) = 1$, $\phi(3) = 2$, $\phi(4) = 2$ (the relevant integers are 1 and 3), $\phi(5) = 4$, $\phi(6) = 2$ (the relevant integers are 1 and 5), and so on. Observe that $\phi(p) = p - 1$ for all primes p.

We now have the following result: *$\phi(n)$ is a power of 2 if and only if a regular n-gon is constructible using a ruler and compass*; in other words, if and only if n is of the form $2^k P_1 \cdots P_r$ where the P_i are distinct Fermat primes.

Examples

(a) For the numbers 3, 5, 6, 10, 15 and 51, we have

$$\phi(3) = 2, \quad \phi(5) = 4, \quad \phi(6) = 2,$$
$$\phi(10) = 4, \quad \phi(15) = 8, \quad \phi(51) = 32;$$

in each case the phi-value is a power of 2.

(b) For the numbers 7, 9, 11, 13, 18 and 21, we have

$$\phi(7) = 6, \quad \phi(9) = 6, \quad \phi(11) = 10,$$
$$\phi(13) = 12, \quad \phi(18) = 6, \quad \phi(21) = 12;$$

none of these phi-values is a power of 2.

11.3 The Euclidean numbers

Recall Euclid's proof for the claim that there are infinitely many primes. The proof uses the number E_n, defined by the equation

$$E_n = p_1 p_2 p_3 \cdots p_n + 1,$$

where p_i denotes the ith prime. The numbers $E_1, E_2, \ldots$ are known as the *Euclidean numbers*. We now take a closer look at these numbers. There are many questions that one may ask about them. Considering their usage in showing that the number of primes is infinite, it may be of interest to check how often these numbers themselves are prime. The results are listed below (Table B.11.2).

After $n = 5$, we have a string of composite numbers; the next prime comes at $n = 11$. Following this, there is a very long string of composites – the next prime comes only at $n = 75$. This prime is a gigantic one – it is a 154-digit number!

Table B.11.2 *Primality status of the Euclidean numbers*

n	E_n	Primality status	Factors if non-prime
1	3	Prime	
2	7	Prime	
3	31	Prime	
4	211	Prime	

Continues

Table B.11.2 *Continued*

n	E_n	Primality status	Factors if non-prime
5	2311	Prime	
6	30031	Non-prime	59×509
7	510511	Non-prime	$19 \times 97 \times 277$
8	9699691	Non-prime	347×27953
9	223092871	Non-prime	317×703763
10	6469693231	Non-prime	$331 \times 571 \times 34231$

The distribution of primes in the sequence is unclear, and the final word on the problem has yet to be written.

Here is an interesting fact about the Euclidean numbers: *No Euclidean number is a power of an integer.*

11.4 Unsolved problems

Anyone who studies the primes deeply is bound to see tantalizing glimpses of some deep order, only to be frustrated by the breakdown of observed patterns. We illustrate this comment with a curiosity, first pointed out by a reader of the problem-solving magazine, *Samasyā*.[3] We first compute the sequence of partial sums of the primes, and then we alternately add 3 and 2 to the sums ($2+3=5$, $5+2=7$, $10+3=13$, and so on). We obtain the following sequence:

5, 7, 13, 19, 31, 43, 61, 79, 103, 131, 163, 199, 241, 283, 331, 383, 443, 503, 571, 641, 715, 793, 877, 965, 1063, 1163, 1267, 1373, 1483, 1595, 1723, 1853, ...

This sequence contains surprisingly many primes. Its first non-prime number is 715, which means that it starts off with 20 primes! Moreover, of the first 100 numbers in the sequence, no less than 52 are prime! Why should this simple procedure generate so many primes? It is hard to say.

⋆ ⋆ ⋆

In the context of the example given above, the reader will not be surprised to hear that there are innumerable problems concerning

[3] See page 42 of *Samasyā*, Volume **4**, Nos. 1, 2 & 3.

the primes that are unsolved as of today. We list a few of them below.

- *Twin primes*
  When two consecutive odd numbers are prime, we refer to the pair as "twin primes".

  Examples

  $$(3,5),\ (5,7),\ (11,13),\ (101,103),\ (107,109).$$

  Are there infinitely many prime twins of this kind? The available evidence suggests that their supply is indeed infinite. Between 1000 and 1500 alone, we find 15 such twins:

  $$\begin{array}{lll} (1019, 1021), & (1031, 1033), & (1049, 1051), \\ (1061, 1063), & (1091, 1093), & (1151, 1153), \\ (1229, 1231), & (1277, 1279), & (1289, 1291), \\ (1301, 1303), & (1319, 1321), & (1427, 1429), \\ (1451, 1453), & (1481, 1483), & (1487, 1489). \end{array}$$

  Here is a pair of particularly large twins:

  $$(1000000000061, 1000000000063).$$

- *Prime triplets*
  Are there infinitely many numbers n such that $(n, n+2, n+6)$ is a "prime triplet", containing only primes?

  Examples

  $$(5,7,11),\quad (11,13,17),\quad (41,43,47).$$

  There are 15 such triplets amongst the numbers below 1000, the largest being $(881, 883, 887)$. A particularly impressive prime triplet is

  $$(1000000005707, 1000000005709, 1000000005713).$$

  The same question holds for the triplet $(n, n+4, n+6)$.

- *Goldbach's conjecture*
  Is it true that every even integer beyond 4 is the sum of two odd primes?

  Examples

  $80 = 7 + 73$, $100 = 3 + 97$, and so on. The available evidence suggests that the answer is: *yes*.

- *Primes of the form* $n^2 + 1$
  Are there infinitely many primes of the form n^2+1 for some integer n?

  Example The primes 17 and 37 are of this form. Indeed, there are 19 such primes below 10 000:

  2, 5, 17, 37, 101, 197, 257, 401, 577, 677, 1297,
  1601, 2917, 3137, 4357, 5477, 7057, 8101, 8837.

  The evidence suggests that there are infinitely many such primes.

- *Mersenne primes*
  Are there infinitely many Mersenne primes?

- *Fermat primes*
  Is there a Fermat prime beyond $F_4 = 2^{16} + 1$?

As Hardy and Littlewood remark, such questions "... may be multiplied, but their proof or disproof is at present beyond the resources of mathematics."

Exercises

11.4.1 Show that the AP

$$2, \ 5, \ 8, \ 11, \ 14, \ \ldots$$

contains infinitely many prime numbers.

11.4.2 Show that the AP

$$6, \ 13, \ 20, \ 27, \ 34, \ \ldots$$

contains infinitely many prime numbers.

11.4.3 Show that of three consecutive numbers, one is necessarily a multiple of 3. (The same is true for three consecutive odd numbers.)

11.4.4 Show that if the numbers a, $a + d$, $a + 2d$ are all prime, with $a > 3$ and $d > 0$, then d is a multiple of 6.

11.4.5 Find examples of three-term APs which contain only primes, other than 3, 13, 23.

11.4.6 Show that if n is composite, then the nth repunit R_n is non-prime.

11.4.7 Show that if $p > 5$ is prime, then p is a divisor of R_{p-1}.

11.4.8 Verify that the numbers 496 and 8128 are perfect.

11.4.9 Show that the Fermat numbers are coprime, and deduce from this that there are infinitely many primes.

11.4.10 Show that a Euclidean number is never a square.

Chapter 12

Polygonal Numbers

We have encountered the triangular numbers and the pentagonal numbers earlier, in Part A. These numbers are *polygonal numbers*, and they are obtained by stacking dots in arrays of different shapes. From triangular arrays we get triangular numbers; from square arrays we get square numbers; from pentagonal arrays we get pentagonal numbers; and so on. For convenience, we will refer to the triangular numbers as T-numbers and to the pentagonal numbers as P-numbers. In this chapter, we present some features of the T- and P-numbers.

12.1 Triangular numbers

The triangular numbers, denoted by T_1, T_2, T_3, $\ldots$, are the numbers

$$1,\ 3,\ 6,\ 10,\ 15,\ 21,\ 28,\ 36,\ 45,\ 55,\ 66,\ 78,\ \ldots.$$

The nth T-number is given by the formula

$$T_n = \frac{n(n+1)}{2},$$

and n is the *rank* of the triangular number T_n. We sometimes in short, we obtain the sequence of squares. This property is easy to explain, for the sum of T_{n-1} and T_n is

$$\frac{(n-1)n}{2} + \frac{n(n+1)}{2} = \frac{n^2 - n + n^2 + n}{2} = \frac{2n^2}{2} = n^2.$$

A pictorial proof – a "proof without words" – may be given by recalling the definition of the T-numbers via triangular dot arrays. The arrays for T_{n-1} and T_n may be joined together as shown below, and it is clear that the two shapes join up to form a square.

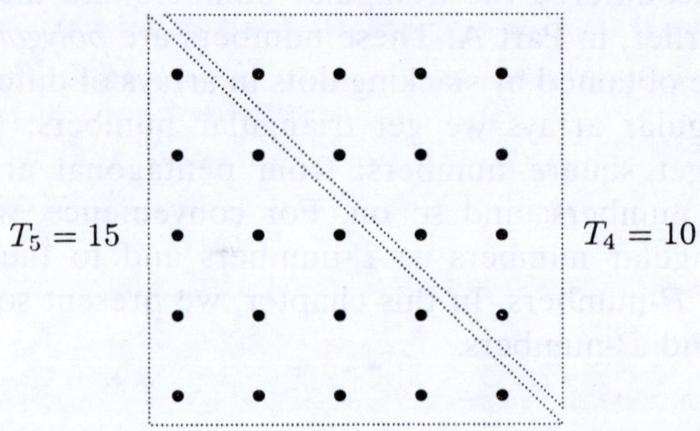

Figure B.12.1 *Illustrating why* $T_5 + T_4 = 5^2$

A still simpler relationship between the T-numbers and the squares is obtained by multiplying the T-numbers by 8 and adding 1 to the result. We obtain the numbers

$$9,\ 25,\ 49,\ 81,\ 121,\ 169,\ 225,\ 289,\ 361,\ \ldots,$$

and these are just the *odd squares*. There is no mystery at all about this connection:

$$8T_n + 1 = 8 \times \frac{n(n+1)}{2} + 1 = 4n^2 + 4n + 1 = (2n+1)^2.$$

Given below (Figure B.12.2.) is a nice visual depiction of this identity. Given 8 copies of the T-number $n(n+1)/2$, we join them up in pairs to form 4 rectangles, each of the size $n \times (n+1)$, and then we stack the rectangles to form a square as shown below. A vacant square is left at the centre of the shape, which we fill with the unused 1. Now, we have a square measuring $(n+1) \times (n+1)$.

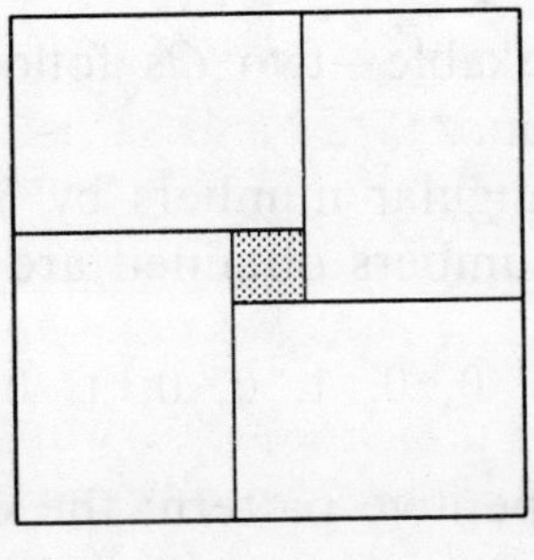

Figure B.12.2 *Illustrating why $8T_n + 1$ is a square*

Conversely, if x is an odd square, then $(x-1)/8$ is a T-number. For example, from the odd square 25, we get the T-number $(169-1)/8 = 21 = T_6$; and from the odd square 841, we get the T-number 105, which is T_{14}. For proof, note that if x is an odd square, then $x = (2n+1)^2 = 4n^2 + 4n + 1$ for some integer n. But now we get

$$\frac{x-1}{8} = \frac{4n^2+4n}{8} = \frac{n(n+1)}{2} = T_n.$$

⋆ ⋆ ⋆

Here is a neat application of the above observation. We shall show that for all non-negative integers, n, the number

$$1 + 9 + 9^2 + \cdots + 9^n$$

is a triangular number. For example, $n = 2$ gives the number $1 + 9 + 81 = 91$ which is T_{13} and $n = 3$ gives the number 820, which is T_{40}. For proof, we write the number in base-9. Let $X = 1 + 9 + 9^2 + \cdots + 9^n$; then in base-9, $X = (111\ldots1)_9$, so

$$8X + 1 = (888\ldots8)_9 + 1 = (1000\ldots0)_9.$$

Thus, $8X + 1$ is a power of 9, and this is a square, because 9 is a square. Therefore, X is a T-number.

⋆ ⋆ ⋆

An interesting pattern in the T-sequence is obtained if we categorize the numbers as odd (O) or even (E). The sequence now looks like this:

$$O,\ O,\ E,\ E,\ O,\ O,\ E,\ E,\ O,\ O,\ E,\ E,\ \ldots$$

The pattern is unmistakable – two *O*s followed by two *E*s, then two *O*s again, and so on.

If we divide the triangular numbers by 3 and record only the remainders, then the numbers obtained are the following:

$$1,\ 0,\ 0,\ 1,\ 0,\ 0,\ 1,\ 0,\ 0,\ 1,\ 0,\ 0,\ \ldots.$$

Here too we find a repetitive pattern; the cycle (1, 0, 0) repeats indefinitely.

If we divide the triangular number by 4, then the remainders we get are the following:

$$1,\ 3,\ 2,\ 2,\ 3,\ 1,\ 0,\ 0,\ 1,\ 3,\ 2,\ 2,\ 3,\ 1,\ 0,\ 0,\ \ldots.$$

In this case, the cycle is (1, 3, 2, 2, 3, 1, 0, 0), and the cycle length is 8.

One may guess at this stage that a repetitive cycle will emerge, independent of which number d we divide the triangular numbers with. This is indeed so. The table displayed below gives the cycles corresponding to various divisors. (For $d = 8$ and $d = 10$, we have not displayed the full cycles, as they are too long.)

Table B.12.1 *Remainders from the sequence of T-numbers*

d	Repeating cycle	Length of cycle
2	(1, 1, 0, 0)	4
3	(1, 0, 0)	3
4	(1, 3, 2, 2, 3, 1, 0, 0)	8
5	(1, 3, 1, 0, 0)	5
6	(1, 3, 0, 4, 3, 3, 4, 0, 3, 1, 0, 0)	12
7	(1, 3, 6, 3, 1, 0, 0)	7
8	(1, 3, 6, 2, . . . , . . . , 3, 1, 0, 0)	16
9	(1, 3, 6, 1, 6, 3, 1, 0, 0)	9
10	(1, 3, 6, 0, . . . , . . . , 3, 1, 0, 0)	20
11	(1, 3, 6, 10, 4, 10, 6, 3, 1, 0, 0)	11

In Table B.12.1, see patterns-within-patterns! For instance, each cycle ends with two 0s. Will this always be the case?

The lengths of the cycles associated with the various divisors are given below.

Divisor (d)	2	3	4	5	6	7	8	9	10	11
Cycle length	4	3	8	5	12	7	16	9	20	11

What is the pattern here? There is something hidden in these numbers, but it is not easy to find! Can you guess the cycle length corresponding to $d = 50$? $d = 100$? $d = 101$?

⋆ ⋆ ⋆

The triangular numbers occur when we consider problems connected with *counting*. Draw n lines on a sheet of paper, making sure that each pair of lines meets somewhere on the sheet. How many points of intersection do we get? The answer is T_{n-1}. (We have considered this problem earlier, in Chapter 6.)

A more fanciful way of saying the same thing is as follows. Let n persons meet at a party, and let everyone shake hands with everyone else exactly once. At the end, how many handshakes have been performed? The answer is T_{n-1}.

Here is a third way in which the triangular numbers arise. Draw an n-sided convex polygon. How many diagonals does the polygon have? The answer is $T_{n-1} - n$. (Naturally, we must have $n \geq 3$ in this problem.)

Each problem may be analyzed in exactly the same way; let us consider the "handshake" problem. Number the guests 1, 2, $\ldots, n-1, n$. Person #1 shakes hands with each of the other guests ($\therefore$ $n-1$ handshakes in all). Person #2 shakes hands with each guest too, but we cannot include the handshake with person #1 as it has already been counted; so we add only $n-2$ to the running total. Person #3 similarly contributes an additional $n-3$ handshakes; and so on. In the end, the total count of handshakes is

$$(n-1) + (n-2) + (n-3) + \cdots + 2 + 1 = T_{n-1}.$$

⋆ ⋆ ⋆

If we compute the cumulative sums of the *reciprocals* of the T-numbers, then another nice pattern emerges. The display below gives some representative values of the sum $\frac{1}{T_1} + \frac{1}{T_2} + \cdots + \frac{1}{T_n}$, which we denote for convenience by a_n. We have $a_1 = 1/1$, and

$$a_2 = \frac{1}{1} + \frac{1}{3} = \frac{4}{3}, \quad a_3 = \frac{4}{3} + \frac{1}{6} = \frac{3}{2}, \quad a_4 = \frac{3}{2} + \frac{1}{10} = \frac{8}{5},$$

$$a_5 = \frac{8}{5} + \frac{1}{15} = \frac{5}{3}, \quad a_6 = \frac{5}{3} + \frac{1}{21} = \frac{12}{7}, \quad \ldots$$

The pattern may not seem obvious, but if we *halve* the numbers $a_1, a_2, a_3, \ldots$, then the pattern jumps out at us. We get the fractions

$$\frac{1}{2}, \frac{2}{3}, \frac{3}{4}, \frac{4}{5}, \frac{5}{6}, \frac{6}{7}, \ldots,$$

and it appears as though we have

$$\frac{a_n}{2} = \frac{n}{n+1}.$$

Is this pattern genuine? Indeed it is, and there is a simple way of proving that this is so. The reader is invited to supply the proof.

According to what we have just found, the following holds good:

$$\frac{1}{T_1} + \frac{1}{T_2} + \cdots + \frac{1}{T_n} = \frac{2n}{n+1}.$$

The fraction of the right side of this equation is

$$\frac{2n}{n+1} = 2 - \frac{2}{n+1},$$

and this quantity gets gradually closer to 2 as n grows larger and larger (because $2/(n+1)$ grows smaller and smaller as n grows). We infer from this that as n grows without bound, the sum of the reciprocals of the first n triangular numbers gets arbitrarily close to 2. Mathematicians express this statement thus: *The sum of the reciprocals of all the triangular numbers equals 2.*

⋆ ⋆ ⋆

The next problem we consider is related to the "coin changing" problem studied in Chapters 7 and 8. Unfortunately, it is difficult to analyze, so we will only quote the result:

> *Every positive integer may be written as the sum of three or fewer triangular numbers.*

Examples $50 = 28 + 21 + 1$; $60 = 36 + 21 + 3$; $70 = 45 + 15 + 10$; $75 = 45 + 15 + 15$; $80 = 55 + 15 + 10$; and so on.

This result was known to Fermat, but the first mathematician to give a clear proof for this was Gauss.

There is a connection between this result and the one on writing integers as sums of squares. Let n be written as a sum of three triangular numbers, say $n = T_a + T_b + T_c$, where a, b, c are non-negative integers. Then, we have

$$n = \frac{a(a+1)}{2} + \frac{b(b+1)}{2} + \frac{c(c+1)}{2}.$$

Multiplying by 8 and adding 3 to both sides, we get

$$8n + 3 = (2a+1)^2 + (2b+1)^2 + (2c+1)^2.$$

So $8n+3$ is a sum of three squares. Conversely, if $8n+3$ is a sum of three squares, then the three squares must all be odd, so n is a sum of three T-numbers.

⋆ ⋆ ⋆

Can a triangular number be a square? This problem too is related to one studied earlier (Chapter 7). Suppose that $T_a = b^2$, where a and b are positive integers. Then, we have

$$8T_a + 1 = 8b^2 + 1, \qquad \therefore\ (2a+1)^2 = 2 \times (2b)^2 + 1.$$

If we write A for $2a+1$ and B for $2b$; then, we get

$$A^2 - 2B^2 = 1.$$

This equation was studied in detail in Chapter 7 (Section 7.6). Its solutions are displayed below:

A	3	17	99	577	3363	...
B	2	12	70	408	2378	...

From the above table, we deduce the following pairs of values for a and b, via $a = (A-1)/2$ and $b = B/2$:

$$(a, b) = (1, 1),\ (8, 6),\ (49, 35),\ (288, 204),\ (1681, 1189),\ \ldots.$$

So T_a is a square when $a = 1, 8, 49, 288, 1681, \ldots$. Conversely, from any solution in positive integers to the equation $A^2 - 2B^2 = 1$, we may obtain a triangular number which is a square.

⋆ ⋆ ⋆

Interestingly, the same connection is revealed in another such question: *When is one triangular number twice another?* Let a and b

be positive integers, such that T_a is twice T_b (example: $T_3 = 2T_2$). Then, after multiplying by 8 and adding 1, we have

$$8T_a + 1 = 2 \times (8T_b + 1) - 1, \quad \therefore \quad (2a+1)^2 = 2(2b+1)^2 - 1.$$

Write $A = 2a + 1$ and $B = 2b + 1$; then $A^2 - 2B^2 = -1$. The solutions of this equation are:

A	1	7	41	239	1393	...
B	1	5	29	169	985	...

From this, we deduce the following pairs of values for a and b, via $a = (A-1)/2$ and $b = (B-1)/2$ (we discard the very first pair, $a = 0$, $b = 0$):

$$(a, b) = (3, 2), \ (20, 14), \ (119, 84), \ (696, 492), \ \ldots.$$

Thus, we have $T_{20} = 210 = 2 \times T_{14}$, $T_{119} = 7140 = 2 \times T_{84}$, and so on.

⋆ ⋆ ⋆

It turns out that no triangular number is a cube or a fourth power; but this is hard to prove!

12.2 Pentagonal numbers

As stated earlier, the pentagonal numbers are obtained when we make pentagonal dot arrays. The sequence of pentagonal numbers $P_1, P_2, P_3, \ldots$ is

$$1, \ 5, \ 12, \ 22, \ 35, \ 51, \ 70, \ 92, \ \ldots.$$

A formula for the nth pentagonal number P_n is easily found:

$$P_n = \frac{n(3n-1)}{2}.$$

We quickly notice that the P-numbers are cumulative sums of the AP

$$1, \ 4, \ 7, \ 10, \ 13, \ 16, \ 19, \ 22, \ \ldots.$$

The questions that had been asked about the T-numbers may be asked here too. For example, (a) *Which pentagonal numbers are squares?* (b) *Can one pentagonal number be twice another?*. We shall not consider these questions here, as the discussion would become repetitive. Instead, we shall make a few remarks about a most extraordinary connection between the pentagonal numbers and the sums of divisors of numbers, first discovered by Euler.

We shall first introduce the 'generalized pentagonal numbers'. These are obtained by substituting negative integers, into the formula $n(3n-1)/2$. Substituting $n = 0, -1, -2, -3, -4, -5, \ldots$ into the formula, we get the numbers

$$0,\ 2,\ 7,\ 15,\ 26,\ 40,\ 57,\ \ldots.$$

Collecting the two sets of numbers together (obtained from the non-negative integers and the negative integers respectively) and arranging them in ascending order, we get

$$0,\ 1,\ 2,\ 5,\ 7,\ 12,\ 15,\ 22,\ 26,\ 35,\ 40,\ 51,\ 57,\ \ldots.$$

These are the *generalized pentagonal numbers*, and they may be obtained from the single formula $n(3n \pm 1)/2$ by substituting $n = 0$, 1, 2, 3, ... For convenience, we shall refer to these numbers as "GP numbers".

Next, for positive integers n define $\sigma(n)$ to be the sum of the divisors of n. ("σ" is a Greek letter and is pronounced "sigma", and $\sigma(n)$ is called – naturally enough! – the sum-of-divisors function.)

Example For $n = 20$, 30 and 40 we have:

$$\begin{aligned}\sigma(20) &= 1+2+4+5+10+20 = 42,\\ \sigma(30) &= 1+2+3+5+6+10+15+30 = 72,\\ \sigma(40) &= 1+2+4+5+8+10+20+40 = 90.\end{aligned}$$

Now, for Euler's discovery!

For any given positive integer n, list all the GP numbers which are less than or equal to n. Suppose that a is the largest such number; then the GP numbers less than or equal to n are 0, 1, 2, 5, 7, 12, 15, 22, ..., a (listed in order). Now, we compute the following sum:

$$\sigma(n) - \sigma(n-1) - \sigma(n-2) + \sigma(n-5) + \sigma(n-7) - \sigma(n-12)$$
$$-\sigma(n-15) + \sigma(n-22) + \sigma(n-26) - - + + \cdots.$$

The last term in the second line is $\sigma(n-a)$. The sequence of signs, after the first term, is $--++--\cdots$. If $a=n$, that is, if n itself is a GP number, then the last term in this expression will be $\sigma(0)$. Now, $\sigma(0)$ does not have any meaning as such, so we must assign a meaning to it. We do so as follows: $\sigma(0)=n$. This is obviously not a fixed meaning—it depends on the value of n. So this is not part of the functional definition of σ; it is merely a convention.

Euler's pentagonal number theorem states the following: *The sum is 0 for every value of n.*

This is an astounding result, and it certainly takes a while to digest. Let us study a few examples to see what is happening behind the scenes.

$\underline{n=10}$ The GP numbers less than or equal to 10 are 0, 1, 2, 5 and 7, so we compute the quantity

$$\sigma(10)-\sigma(9)-\sigma(8)+\sigma(5)+\sigma(3).$$

Now $\sigma(10)=18$, $\sigma(9)=13$, $\sigma(8)=15$, $\sigma(5)=6$ and $\sigma(3)=4$ (please check these values!); so the sum on the left side is

$$18-13-15+6+4=0,$$

as predicted.

$\underline{n=20}$ The GP numbers less than or equal to 20 are 0, 1, 2, 5, 7, 12 and 15, so the sum to be computed is

$$\sigma(20)-\sigma(19)-\sigma(18)+\sigma(15)+\sigma(13)-\sigma(8)-\sigma(5).$$

Now $\sigma(20)=42$, $\sigma(19)=20$, $\sigma(18)=39$, $\sigma(15)=24$, $\sigma(13)=14$, $\sigma(8)=15$, $\sigma(5)=6$; so the sum on the left side is

$$42-20-39+24+14-15-6=0,$$

as predicted.

$\underline{n=22}$ The GP numbers less than or equal to 22 are 0, 1, 2, 5, 7, 12, 15 and 22, so the sum to be computed is

$$\sigma(22)-\sigma(21)-\sigma(20)+\sigma(17)+\sigma(15)-\sigma(10)-\sigma(7)+\sigma(0).$$

This time we find a $\sigma(0)$ in the summation, so we assign to it a value of 22, because $n=22$ here. The other values are: $\sigma(22)=36$,

$\sigma(21) = 32$, $\sigma(20) = 42$, $\sigma(17) = 18$, $\sigma(15) = 24$, $\sigma(10) = 18$, $\sigma(7) = 8$. The sum to be computed is therefore,

$$36 - 32 - 42 + 18 + 24 - 18 - 8 + 22 = 0,$$

as it ought to be.

Euler's law has triumphed every time!

⋆ ⋆ ⋆

Another such discovery, just as astounding, was made by the mathematician Carl Jacobi; it finds a connection between the triangular numbers and the pentagonal numbers. To understand the connection, we start with the following expression in x:

$$1 - x - x^2 + x^5 + x^7 - x^{12} - x^{15} + x^{22} + x^{26} - \cdots.$$

This is an unending expression in powers of x, and it is called a *power series* in x. Loosely speaking, we may think of it as a polynomial of infinite degree. Observe that the exponents are precisely the GP numbers, and the signs follow the same pattern as in Euler's theorem: $- - + + - - \cdots$.

Jacobi's plan is to find the *cube* of this expression. Now the expression is an infinite one, so how are we to compute its cube? We may do so as follows. We chop off all terms beyond some point, then cube the resulting expression (which is now just a polynomial), and then from the answer we chop off all terms that would have been affected by the omission of terms in the original power series. Thus, we may start with the polynomial $1 - x$; its cube is $1 - 3x + 3x^2 - x^3$, but we must delete all terms with degree greater than 1. We are left with the expression $1 - 3x$.

If we start with the polynomial $1 - x - x^2$, we find that it's cube is

$$1 - 3x + 5x^3 - \cdots$$

(we have deleted all terms with degree ≥ 5). If we start with the polynomial $1 - x - x^2 + x^5$, we find its cube is

$$1 - 3x + 5x^3 - 7x^6 + \cdots$$

(this time all terms with degree ≥ 7 have been omitted); if we start with $1 - x - x^2 + x^5 + x^7$, we find its cube to be

$$1 - 3x + 5x^3 - 7x^6 + 9x^{10} - \cdots$$

(all terms with degree ≥ 12 have been omitted); if we start with the polynomial $1 - x - x^2 + x^5 + x^7 - x^{12}$, we find its cube to be

$$1 - 3x + 5x^3 - 7x^6 + 9x^{10} - \cdots$$

(the same as in the preceding case); and the cube of the polynomial $1 - x - x^2 + x^5 + x^7 - x^{12} - x^{15}$ is

$$1 - 3x + 5x^3 - 7x^6 + 9x^{10} - 11x^{15} + 13x^{21} - \cdots.$$

(All terms with degree ≥ 22 have been deleted.) No one can miss the pattern in these answers! Here is Jacobi's theorem:

$$\begin{aligned}&\left(1 - x - x^2 + x^5 + x^7 - x^{12} - x^{15} + x^{22} + x^{26} - \cdots\right)^3\\&= 1 - 3x + 5x^3 - 7x^6 + 9x^{10} - 11x^{15} + 13x^{21} - \cdots.\end{aligned}$$

The exponents are the triangular numbers, and the coefficients are the odd numbers, with alternating signs. What an extraordinary result!

Exercises

For the sake of convenience and easy reference, we give below a list of values of $\sigma(n)$ for values of n between 1 and 100.

12.2.1 Find pairs of triangular numbers which bear the ratio 1 : 3 to one another.

12.2.2 Observe that 3 is a Fermat number as well as a triangular number. Are there other numbers which belong to both sequences?

12.2.3 Show that

$$\frac{1}{T_1} + \frac{1}{T_2} + \cdots + \frac{1}{T_n} = \frac{2n}{n+1}.$$

LIST OF VALUES OF $\sigma(n)$ FOR $1 \leq n \leq 100$

	1	2	3	4	5	6	7	8	9	10
0+	1	3	4	7	6	12	8	15	13	18
10+	12	28	14	24	24	31	18	39	20	42
20+	32	36	24	60	31	42	40	56	30	72
30+	32	63	48	54	48	91	38	60	56	90
40+	42	96	44	84	78	72	48	124	57	93
50+	72	98	54	120	72	120	80	90	60	168
60+	62	96	104	127	84	144	68	126	96	144
70+	72	195	74	114	124	140	96	168	80	186
80+	121	126	84	224	108	132	120	180	90	234
90+	112	168	128	144	120	252	98	171	156	217

The array is to be read as follows. To find $\sigma(45)$, we look up the row labelled "40+" and the column labelled "5". The entry in the appropriate box is 78, $\therefore \sigma(45) = 78$.

To find $\sigma(72)$, we look up the row labelled "70 +" and the column labelled "2". The entry at the intersection of this row and column is 195, $\therefore \sigma(72) = 195$.

Further examples are $\sigma(25) = 31$, $\sigma(36) = 91$, $\sigma(92) = 168$.

The table has been prepared using the following formula: if the prime factorization of n is $n = p^a \times q^b \times \cdots$, then

$$\sigma(n) = \frac{p^{a+1} - 1}{p - 1} \times \frac{q^{b+1} - 1}{q - 1} \times \cdots.$$

For example, let $n = 45 = 3^2 \times 5^1$; then we get

$$\sigma(45) = \frac{3^3 - 1}{3 - 1} \times \frac{5^2 - 1}{5 - 1} = 13 \times 6 = 78,$$

which is in agreement with the value obtained earlier.

Chapter 13

Egyptian Fractions

This chapter represents a change in tone – the topic we shall be studying is not going to be heavily number theoretic, as the past few chapters undoubtedly have been. The topic we shall study is *unit fractions* or *Egyptian fractions* – fractions which, in reduced form, have a numerator of 1:

$$\frac{1}{1}, \frac{1}{2}, \frac{1}{3}, \frac{1}{4}, \frac{1}{5}, \frac{1}{6}, \ldots.$$

It is a curious fact that in an earlier era, the Egyptians used to represent numbers as sums of unit fractions. In the "Rhind Papyrus", excavated from the ruins of a building in Egypt and dating from ancient times (4000 BC or so), we find clear evidence of this. The document has, among other things, a division table giving the values of $2/n$ for all odd n from 3 to 101, and another table giving the values of $n/10$ for all n from 1 to 9. The answers are invariably expressed in terms of unit fractions, with one exception – the fraction 2/3, which they accepted as a fraction, "as is". Thus, 4/10 is written as $1/3+1/15$, and 7/10 as $2/3+1/30$. These fractions are not expressed the way we do now; rather, there is a special symbol for each such fraction ($1/n$ is written by placing a dot above the symbol for n). Clearly, any fraction can be written as a sum of unit fractions; for example, 3/7 may be written as $1/7+1/7+1/7$. But the Egyptians avoided such representations; their choice in such cases would have been $3/7 = 1/3 + 1/11 + 1/231$. Why they chose to do this, however, is not too clear.

⋆ ⋆ ⋆

When a list of numbers $a, b, c, d, \ldots$ forms an AP, their reciprocals,

$$\frac{1}{a}, \frac{1}{b}, \frac{1}{c}, \frac{1}{d}, \ldots$$

are said to form a *harmonic progression* (often shortened to HP). Since the numbers 1, 2, 3, 4, 5, ... lie in an AP, the unit fractions lie in a harmonic progression. For this reason, the unit fractions are also known as the *harmonic numbers*, and the infinite series

$$1 + \frac{1}{2} + \frac{1}{3} + \frac{1}{4} + \frac{1}{5} + \cdots$$

is often referred to as the *harmonic series*.

⋆ ⋆ ⋆

There is a reason for the use of the word "harmonic", and it goes back to the time of Pythagoras, whose school discovered the mathematical basis of the musical scale. They knew, for instance, the "law of small numbers" in music – the law that notes sound good when played together only if the ratios of their frequencies can be expressed using small numbers. For example, two notes may be in unison (frequency ratio 1 : 1); or they may be an octave apart (frequency ratio 2 : 1); or three notes may form a 'fifth' (in the Indian system this is the scale SA–PA–$\overline{\text{SA}}$, where the second SA, with a line over it, is one octave higher than the first one; here the frequencies are in the ratios 2 : 3 : 4); and so on.

13.1 Divergence of the harmonic series

The first result that we shall prove is a surprising one – it runs quite contrary to intuition! This is the statement that the sum

$$1 + \frac{1}{2} + \frac{1}{3} + \frac{1}{4} + \frac{1}{5} + \cdots$$

grows without bound as we successively add more terms. In other words, by adding sufficiently many terms we can make the sum exceed any given number. The mathematician describes this by saying that *the harmonic series diverges*.

As stated earlier, the result is counter-intuitive. It is useful to check first what the numerical evidence suggests. Let $S(n)$ denote the sum of the first n unit fractions:

$$S(n) = 1 + \frac{1}{2} + \frac{1}{3} + \cdots + \frac{1}{n}.$$

Using a computer, we get the following table of values:

n	10	100	1000	10000	100000
$S(n)$	2.93	5.19	7.49	9.79	12.09

The value of $S(n)$ does seem to increasing steadily—but very very slowly! How can this divergence be shown?

⋆ ⋆ ⋆

A few remarks about the concepts of limit and convergence and divergence of an infinite series need to be made at the start.

Let $a_1, a_2, a_3, \ldots$, be a sequence of numbers. We say that the sequence *converges* (or "tends") to a number L if, as n grows without bound, a_n gets arbitrarily close to L. More precisely, we say that the sequence tends to L if, given any positive quantity ϵ, no matter how tiny, from some point onwards all the a_is lie within a distance of ϵ from L.

The following statements are different ways of saying the same thing: given a sequence a_1, a_2, a_3, … and a number L,

- the sequence tends to L;
- $a_n \to L$ as n grows without bound (*Note* The phrase "n grows without bound" is usually written in short form as $n \to \infty$. Here '∞' is the symbol for infinity. Note that infinity is merely a synonym for "an indefinitely large number".);
- the limit of $\{a_n\}$ is L; or $\lim a_n = L$.

Given a sequence of numbers $a_1, a_2, a_3, \ldots$, let us consider what meaning might be given to the "infinite sum" $a_1 + a_2 + a_3 + \cdots$. Such a sum is termed an *infinite series*, or simply "series" for short. Let a sequence $\{b_n\}$ be defined thus:

$$\begin{aligned} b_1 &= a_1, \\ b_2 &= a_1 + a_2, \\ b_3 &= a_1 + a_2 + a_3, \end{aligned}$$

and so on; b_n is the sum $a_1 + a_2 + \cdots + a_n$. The bs are called the *partial sums* of the as. If the partial sums tend to a limit S (i.e., we have $b_n \to S$ as $n \to \infty$), then we say: *The series* $a_1 + a_2 + a_3 + \cdots$, *converges to* S.

Other outcomes are also possible. The partial sums may grow without bound ($b_n \to \infty$ or $-\infty$ as $n \to \infty$); in this case, we say that the series $a_1 + a_2 + \cdots$ *diverges*. Or the partial sums may jump about without zeroing in on any number (i.e., the bs have no limit). In this case, we say that the series *fails to converge*. We give a few examples below to illustrate the meanings of these terms.

(a) The series $1+\frac{1}{2}+\frac{1}{4}+\cdots+\frac{1}{2^n}+\cdots$ (the sum of the reciprocals of the powers of 2) converges to 2.

(b) The series $1+\frac{1}{3}+\frac{1}{9}+\cdots+\frac{1}{3^n}+\cdots$ (the sum of the reciprocals of the powers of 3) converges to 3/2.

(c) The series $1+\frac{1}{10}+\frac{1}{100}+\cdots+\frac{1}{10^n}+\cdots$ which corresponds to the decimal number 0.1111..., converges to 1/9.

(d) The series $1+1+1+\cdots$ diverges (rather obviously).

(e) The series $1-1+1-1+1-1+1-\cdots$ fails to converge, because the partial sums are alternately 0 and 1 and do not possess a limit.

We shall now show that the harmonic series diverges. The proof given below is due to the Frenchman Nicolo Oresme and it dates to 1350 or so. Historically, it provides the first instance of a series being shown to diverge.

We note the following sequence of inequalities:

$$\frac{1}{3}+\frac{1}{4} > \frac{1}{2},$$

$$\frac{1}{5}+\frac{1}{6}+\frac{1}{7}+\frac{1}{8} > \frac{1}{2},$$

$$\frac{1}{9}+\frac{1}{10}+\cdots+\frac{1}{15}+\frac{1}{16} > \frac{1}{2},$$

$$\frac{1}{17}+\frac{1}{18}+\frac{1}{19}+\cdots+\frac{1}{30}+\frac{1}{31}+\frac{1}{32} > \frac{1}{2},$$

and so on. We see that consecutive terms of the harmonic series can be grouped together so that, for each group, the sum exceeds 1/2. Since there are infinitely many groups, it follows that the sum exceeds any bound that we may set; that is, the sum is infinite.

And that's all there is to the proof!

There is more that can be said. By noting how the terms were grouped above, we can deduce the following, more precise statement:

$$S(2^n) > 1 + \frac{n}{2}.$$

This means that by choosing n large enough, $S(2^n)$ may be made to exceed any given bound. For instance, if we wanted the sum to exceed 1000, then a mere 2^{1998} terms would suffice! This suggests the extreme slowness of the growth of $S(n)$ with n. Nevertheless, $S(n)$ grows without bound, and we may say, rather loosely, that $S(\infty) = \infty$.

★ ★ ★

In the light of the fact that the harmonic series diverges, the result to follow comes as a second surprise. Suppose that from the set of all positive integers we delete all those which have a '0' in their base-10 representation. So we delete the integers 10, 20, 30, ..., 90, 100, 101, 102, ... The integers left now are 1, 2, ..., 8, 9, 11, 12, ..., 18, 19, 21, 22, ..., 28, 29, 31, 32, ... What is the sum of the reciprocals of *this* set of integers? Is it infinite too? The answer (surprise!) is: *no*. (Indeed, the sum does not even exceed 25.)

Though it is not hard to show this, the argument does require some dexterity with algebra. First, some notation: we denote by A_k the set of all k-digit positive integers with no 0s (all references to digital representation are to the base-10 form). Thus,

$$\begin{aligned} A_1 &= \{1, 2, 3, 4, 5, 6, 7, 8, 9\}, \\ A_2 &= \{11, 12, \ldots, 18, 19, 21, 22, \ldots, 99\}, \\ A_3 &= \{111, 112, \ldots, 118, 119, 121, \ldots, 999\}, \end{aligned}$$

and so on. Observe that A_1 has 9 numbers, A_2 has $9^2 = 81$ numbers, A_3 has $9^3 = 729$ numbers, ... Next, we denote by s_k the sum of the reciprocals of the numbers in A_k, and by S the sum of the reciprocals of all the numbers with no 0s in their base-10 representation; then $S = s_1 + s_2 + \ldots$ We wish to show that S is finite.

Now, the numbers in A_1 all lie between 1 and 10; those in A_2 all lie between 10 and 100; those in A_3 all lie between 100 and 1000; those in A_4 all lie between 1000 and 10000; and so on. More generally, we may say that all the numbers in A_k lie strictly between 10^{k-1} and 10^k. So the reciprocals of the numbers in A_k lie strictly between $1/10^k$ and $1/10^{k-1}$, and this is so for each k:

$$\frac{1}{10^k} < \frac{1}{n} < \frac{1}{10^{k-1}} \quad \text{for all } n \in A_k.$$

We now sum up all such inequalities corresponding to the integers n in A_k. Since A_k has 9^k members, there are 9^k such inequalities, and we obtain the following result:

$$\frac{9^k}{10^k} < s_k < \frac{9^k}{10^{k-1}}.$$

This may be written more conveniently thus:

$$0.9^k < s_k < 10 \times 0.9^k \quad \text{for } k = 1, 2, \ldots.$$

We have succeeded in bounding s_k from both sides—a crucial step. The bounds may look absurdly extravagant, but in fact they are good enough for our purposes. They imply, by adding over all values of k, that

$$(0.9 + 0.9^2 + \ldots) < S < 10 \times \left(0.9 + 0.9^2 + \cdots\right).$$

The bracketed series $0.9 + 0.9^2 + \cdots$, which occurs on both sides, is easily summed (see Problem 13.2.4); the sum is

$$\frac{0.9}{1 - 0.9} = \frac{0.9}{0.1} = 9.$$

We see therefore, that S lies between 9 and $10 \times 9 = 90$. So S is certainly a finite number.

Using a computer, the value of S may be estimated more accurately. The computer yields the following values for the s_k (the sum of the reciprocals of all numbers in A_k):

k	s_k
1	2.828968
2	2.065513
3	1.824253
4	1.638774
5	1.474633
6	1.32715

On examining the figures, we find that if we temporarily leave out the numbers s_1 and s_2, then the numbers s_3, s_4, s_5, s_6, ..., *almost* form a geometric progression, with a common ratio close to 0.9:

$$\frac{s_4}{s_3} \approx 0.898, \quad \frac{s_5}{s_4} \approx 0.8998, \quad \frac{s_6}{s_5} \approx 0.89999.$$

Once we noticed this, it does not seem too surprising; after all, there are 9 times as many numbers in A_{k+1} as in A_k, but the numbers in A_{k+1} are about 1/10 as large as the numbers in A_k. So we may guess that the sum $S = s_1 + s_2 + s_3 + s_4 + \cdots$, is roughly given by

$$\begin{aligned} S &\approx 2.828968 + 2.065513 + 1.824253 \\ &\quad + 1.638774(1 + 0.9 + 0.9^2 + 0.9^3 + \cdots) \\ &= 6.71873 + (1.638774 \times 10) \\ &= 6.71873 + 16.38774 = 23.1065. \end{aligned}$$

The actual sum S is certainly *less* than this quantity, but not by much, and we may guess that $S \approx 23.1$.

⋆ ⋆ ⋆

There are many sets of positive integers for which finding the sum of the reciprocals of the members of the set represents an interesting (and possibly difficult) problem. We saw earlier (Section 12.1) that for the set of triangular numbers $\{1, 3, 6, 10, \ldots\}$, the sum of the reciprocals is 2:

$$\frac{1}{1} + \frac{1}{3} + \frac{1}{6} + \frac{1}{10} + \ldots = 2.$$

We may show that for the set of pentagonal numbers $\{1, 5, 12, 22, \ldots, \}$, the sum of the reciprocals is approximately 0.741. This is not a "clean" whole number; indeed, it is not even rational. The exact sum, computed in terms of known constants and by using more advanced methods, is

$$\frac{3 \ln 3}{2} - \frac{\sqrt{3}\pi}{6} \approx 0.741018750885.$$

Here, $\ln 3$ refers to the natural logarithm of 3, and π is, of course, the constant 3.14159.... Curious!

⋆ ⋆ ⋆

A particularly interesting problem arises when we try to sum the reciprocals of the squares:

$$1+\frac{1}{2^2}+\frac{1}{3^2}+\frac{1}{4^2}+\cdots.$$

It is not too hard to show that the series converges: $(k-1)k < k^2 < k(k+1)$ for each positive integer k, so

$$\frac{1}{k(k+1)} < \frac{1}{k^2} < \frac{1}{(k-1)k}, \quad \text{for } k=2,3,\ldots.$$

Now, we have

$$\frac{1}{k(k+1)} = \frac{1}{k} - \frac{1}{k+1},$$

as may be checked (this is the so-called "partial fraction" decomposition of $1/k(k+1)$), and similarly for the term $1/k(k-1)$; so we get

$$\frac{1}{k} - \frac{1}{k+1} < \frac{1}{k^2} < \frac{1}{k-1} - \frac{1}{k} \quad \text{for } k=2,3,\ldots.$$

If we sum these inequalities for $k = 2, 3, \ldots, n$, a lot of cancellation of fractions takes place, and we get

$$\frac{1}{2} - \frac{1}{n+1} < \frac{1}{2^2} + \frac{1}{3^2} + \cdots + \frac{1}{n^2} < 1 - \frac{1}{n}.$$

Let a_n denote the partial sum $1+\frac{1}{2^2}+\frac{1}{3^2}+\cdots+\frac{1}{n^2}$. Then, we have

$$\frac{3}{2} - \frac{1}{n+1} < a_n < 2 - \frac{1}{n}.$$

This shows that a_n never exceeds 2. Since the as steadily increase, yet never exceed 2, we may surmise that $a_n \to L$ for some number L. This seems intuitively true, and it is indeed true; but to show this more rigorously, we must appeal to more advanced methods. We shall assume in what follows that there does exist a limiting value L, and we proceed now to estimate its value.

Write b_n for the sum $1/(n+1)^2 + 1/(n+2)^2 + \cdots$ (this is an infinite series); then $L = a_n + b_n$ for each n. We use once again, the inequality

$$\frac{1}{k} - \frac{1}{k+1} < \frac{1}{k^2} < \frac{1}{k-1} - \frac{1}{k}.$$

Summing this for $k = n+1,\ n+2,\ \ldots$, we get (after lots of cancellation; in fact, an infinite amount of it!):

$$\frac{1}{n+1} < b_n < \frac{1}{n}.$$

So we have the following, extremely useful result: for each n,

$$a_n + \frac{1}{n+1} \ < \ L \ < \ a_n + \frac{1}{n}.$$

Using the computer, we find the following:

$$a_{10} = 1.549677, \qquad a_{100} = 1.634984,$$
$$a_{1000} = 1.643934, \quad a_{10000} = 1.644834.$$

The values of a_{100}, a_{1000} and a_{10000} yield the following, successively better estimates for the sum L:

$$1.64488489 < L < 1.64498489,$$
$$1.64493356 < L < 1.64493456,$$
$$1.64493406 < L < 1.64493407.$$

We are getting close! It appears, then, that $L \approx 1.644934$.

Of course, all we have done is to *estimate* L; we still do not know just what number it is. Mathematicians are often more interested in knowing a number *exactly* – in terms of known constants, for example – than in estimating it. So the question of knowing the exact limit remains. In the 18th century this question tantalized and exasperated many mathematicians, until finally the great Leonhard Euler found the answer. He showed, after a wonderful and highly unusual piece of reasoning, that L can be expressed in terms of the familiar number π by the formula $L = \pi^2/6$. That is,

$$1 + \frac{1}{2^2} + \frac{1}{3^2} + \frac{1}{4^2} + \cdots = \frac{\pi^2}{6}.$$

13.2 An Egyptian algorithm

Can every rational number r between 0 and 1 be written as a sum of distinct unit fractions? The answer is *yes* – and there are infinitely many ways of doing so! The infinity of ways need not surprise us, because we can use the identity

$$\frac{1}{n} = \frac{1}{n+1} + \frac{1}{n(n+1)}$$

to break up the smallest term in a representation into two terms; and then we can do this to the smallest term in the *new* representation; and so on. For example, from the representation for 3/5 as

$$\frac{3}{5} = \frac{1}{2} + \frac{1}{10}$$

we obtain, successively,

$$\begin{aligned}\frac{3}{5} &= \frac{1}{2} + \frac{1}{10} \\ &= \frac{1}{2} + \frac{1}{11} + \frac{1}{110} \\ &= \frac{1}{2} + \frac{1}{11} + \frac{1}{111} + \frac{1}{12210} \\ &= \frac{1}{2} + \frac{1}{11} + \frac{1}{111} + \frac{1}{12211} + \frac{1}{149096310},\end{aligned}$$

(here 149096310 is 12210×12211), and so on. Clearly, this can be continued indefinitely.

So from a single representation, one may obtain infinitely many representations of a given rational number as a sum of distinct unit fractions. The question therefore, is: how do we obtain the initial representation?

Fortunately, this is not hard to do; indeed, there is an elegant and simple algorithm which yields the required representation. Because of its central idea, it is termed a "greedy algorithm". The ancient Egyptians seem to have used this algorithm in their computations.

We simply find the largest unit fraction $1/a$, which is less than or equal to the given rational number r (remember that r lies between 0 and 1), and write $r' = r - 1/a$. If $r' = 0$, then we are through; there is nothing more to be done (we get $r = 1/a$).

If $r' > 0$, then we find the largest unit fraction $1/b$ less than or equal to r', and write $r'' = r' - 1/b$. We shall show below that $b > a$, that is, that a and b are *distinct*.

If $r'' = 0$, there is nothing more to be done (we get $r = 1/a + 1/b$). If $r'' > 0$, then we find the largest unit fraction $1/c$ less than or equal to r'' (we will have $c > b$, just as we had $b > a$); and so we continue .... As will be shown shortly, the sequence of steps cannot go on forever. In the end, we will have a representation for r as a sum of distinct unit fractions:

$$r = \frac{1}{a} + \frac{1}{b} + \frac{1}{c} + \cdots .$$

A program in BASIC may be written to execute the steps. In the program displayed below, the given number r is entered into the 3rd line of the program (as "r = 7/9", or whatever). The output is a list of integers—these are the denominators of the various fractions occurring in the representation.

```
CLS
REM Egyptian fractions
r = 7 / 9
x = r
q = 1 / x
WHILE q > INT(q)
    a = INT(q) + 1: PRINT a
    x = x - 1 / a
    q = 1 / x
WEND
```

For $r = 7/9$, the program outputs the integers 2, 3 and 42. This means, of course, that $7/9 = 1/2 + 1/3 + 1/42$.

The best way to understand how the algorithm functions is to study several examples.

(a) Let $r = 3/5$. The largest unit fraction less than or equal to $3/5$ is $1/2$; so $a = 2$ and $r' = 3/5 - 1/2 = 1/10$, which is a unit fraction. Therefore, we have,

$$\frac{3}{5} = \frac{1}{2} + \frac{1}{10}.$$

(b) Let $r = 7/9$. As earlier, $a = 2$, and $r' = 7/9 - 1/2 = 5/14$. The largest unit fraction less than or equal to r' is $1/3$, and $r'' = 5/14 - 1/3 = 1/42$, which is a unit fraction. Therefore,

$$\frac{7}{9} = \frac{1}{2} + \frac{1}{3} + \frac{1}{42}.$$

(c) Let $r = 10/13$. This time, $a = 2$ and $r' = 10/13 - 1/2 = 7/26$. Next, $b = 4$ and $r'' = 7/26 - 1/4 = 1/52$. Therefore,

$$\frac{10}{13} = \frac{1}{2} + \frac{1}{4} + \frac{1}{52}.$$

(d) Let $r = 11/29$. Then $a = 3$, $r' = 11/29 - 1/3 = 4/87$, $b = 22$, $r'' = 4/87 - 1/22 = 1/1914$, and so

$$\frac{11}{29} = \frac{1}{3} + \frac{1}{22} + \frac{1}{1914}.$$

Occasionally, the greedy algorithm produces fractions with large denominators. For instance, if $r = 7/23$, then the algorithm yields the following:

$$\frac{7}{23} = \frac{1}{4} + \frac{1}{19} + \frac{1}{583} + \frac{1}{1019084},$$

and for $r = 999/1000$, it yields

$$\frac{999}{1000} = \frac{1}{2} + \frac{1}{3} + \frac{1}{7} + \frac{1}{44} + \frac{1}{12158} + \frac{1}{1404249000}.$$

But sometimes the denominators are *very* large indeed; for instance, for $r = 5/121$, the algorithm yields the following:

$$\frac{5}{121} = \frac{1}{25} + \frac{1}{757} + \frac{1}{763309} + \frac{1}{873960180913} + \frac{1}{1527612795642093418846225}.$$

A still more impressive instance of this kind is seen for the fraction 65/131. Here, the largest denominator obtained is a number with 170 digits!

⋆ ⋆ ⋆

We still have to show that the Egyptian algorithm really works; that is, that it comes to a halt after a finite number of steps and yields the required representation. But this is not hard to do.

Let r be the given rational number ($0 < r < 1$), say $r = m/n$, where m and n are coprime positive integers. If r is itself a unit fraction ($m = 1$) there is nothing further to be done, so we assume that it is not. Let $1/a$ be the largest unit fraction less than r; then $a > 1$, and

$$\frac{1}{a} < \frac{m}{n} < \frac{1}{a-1}.$$

Cross-multiplying to clear the fractions, we deduce that

$$ma - n > 0, \qquad m(a-1) - n < 0.$$

This implies that $0 < ma - n < m$; that is, $ma - n$ is an integer lying strictly between 0 and m. Now, according to the scheme, r is replaced by r' where

$$r' = r - \frac{1}{a} = \frac{m}{n} - \frac{1}{a} = \frac{ma-n}{na},$$

and the next unit fraction $1/b$ is obtained from the inequalities $1/b \le r' < 1/(b-1)$, that is

$$\frac{1}{b} \le \frac{m}{n} - \frac{1}{a} \le \frac{1}{b-1}.$$

After transposition of terms, this yields $1/a + 1/b \le m/n$. Since we also have $m/n < 1/(a-1)$, we get $1/a + 1/b < 1/(a-1)$, giving

$$\frac{1}{b} < \frac{1}{a-1} - \frac{1}{a}, \quad \text{or} \quad b > a(a-1).$$

Since $a > 1$, the last result yields $b > a$. In the same way we get $c > b$, and so on; so $a, b, c, \ldots$ are distinct positive integers.

The numerator of r' is either $ma - n$ or a smaller number (this will be the case if some cancellation takes place between $ma - n$ and na; but, in fact, no such cancellation can take place. Do you see why?). Since $ma - n < m$ and $r' > 0$, the numerator of r' is positive but strictly less than that of r.

Arguing in the same way, we see that the numerator of r'' is strictly less than that of r'; the numerator of r''' is strictly less than that of r''; and so on. Each numerator is positive, and the numerators steadily decrease, so sooner or later some numerator must touch 1. At this point, we reach a unit fraction and we can stop our calculations; the algorithm has done its job.

We see, in this manner, that the algorithm always works.

Exercises

13.2.1 Show that $S(2^n) < n + 1$. (Thus, $S(2^n)$ lies between $n/2$ and $n + 1$.)

13.2.2 Show that the sum of the infinite series

$$1+\frac{1}{2}+\frac{1}{4}+\frac{1}{8}+\cdots+\frac{1}{2^n}+\cdots$$

equals 2. (First, use a computer to convince yourself that the result seems to be true!)

13.2.3 Show, similarly, that the sum of the infinite series

$$1+\frac{1}{3}+\frac{1}{9}+\frac{1}{27}+\cdots+\frac{1}{3^n}+\cdots$$

equals 3/2.

13.2.4 Show that the sum of the infinite series

$$1+0.9+0.9^2+0.9^3+0.9^4+\cdots$$

equals 10.

13.2.5 In Section 13.1, we found that s_{k+1}/s_k is nearly 0.9, but not quite; in each case the ratio is slightly less than 0.9. How can this be explained?

13.2.6 Show that the sum of the reciprocals of the integers which have no 1s in their base-10 representation is finite; and estimate the sum, using a computer where necessary.

13.2.7 Using a computer, estimate the sum of the reciprocals of the integers which in base-10 are made up only of 1s.

13.2.8 Using a computer, estimate the sum of the reciprocals of the integers which in base-10 are made up only of 1s and 2s.

Chapter 14

Square Roots

In this chapter we move further away from the integers! In the last chapter we had made a foray into the world of the rationals; now, we step into the land of the irrationals. The subject of study here will be the sequence of square roots of the positive integers, that is, the numbers

$$1,\ \sqrt{2},\ \sqrt{3},\ \sqrt{4}=2,\ \sqrt{5},\ \sqrt{6},\ \sqrt{7},\ \sqrt{8},\ \sqrt{9}=3,\ \sqrt{10},\ \ldots.$$

Expressed as decimal fractions, rounded-off to 3 d.p., the numbers are the following (the last number shown is $\sqrt{20}$):

$$\begin{aligned}&1,\ 1.414,\ 1.732,\ 2,\ 2.236,\ 2.449,\ 2.646,\ 2.828,\ 3,\\ &3.162,\ 3.317,\ 3.464,\ 3.606,\ 3.742,\ 3.873,\ 4,\ 4.123,\\ &4.243,\ 4.359,\ 4.472,\ \ldots\end{aligned}$$

If we strip away the decimal portions of these numbers, leaving only the integer parts, we obtain:

$$1,\ 1,\ 1,\ 2,\ 2,\ 2,\ 2,\ 2,\ 3,\ 3,\ 3,\ 3,\ 3,\ 3,\ 3,\ \ldots$$

The sequence has three 1s, five 2s, seven 3s, nine 4s, . . . ; a pretty pattern, but does it come as a surprise? Not quite – it follows easily from the identity

$$(n+1)^2 - n^2 = 2n+1,$$

which implies that the integers k for which the integer part of $\sqrt{k}$ is n are n^2, n^2+1, n^2+2, . . . , n^2+2n, and these are $2n+1$ in number.

More striking is the sequence of decimal parts. Here we strip away the integer part, retaining only the portion *after* the decimal point. We obtain the following:

0, 0.414, 0.732, 0, 0.236, 0.449, 0.646, 0.828, 0,
0.162, 0.317, 0.464, 0.606, 0.742, 0.873, 0, 0.123,
0.243, 0.359, 0.472, ...

Does this sequence possess a pattern? It certainly does not seem to possess any pattern at all. Is it, then, to be considered as a "random" sequence? This is a difficult and tricky question, and we shall make a few remarks about it later on in the chapter.

★ ★ ★

A conspicuous property of the sequence of square roots is that each member of the sequence is either an integer or else it is irrational. We have encountered irrational numbers earlier – in the "Squares" chapter we showed that $\sqrt{2}$ is irrational. We shall not dwell on this matter here, except to say that for prime numbers p, the irrationality of $\sqrt{p}$ may be proved along the same lines as that of $\sqrt{2}$. (See Problems 14.4.1 and 14.4.2.)

14.1 Closeness

In the sequence of squares 1, 4, 9, 16, 25, ..., the gap between adjacent members grows steadily, and this is true for cubes and fourth powers too. In the sequence of natural numbers 1, 2, 3, 4, 5, ..., the gap between adjacent members stays constant. In the sequence of square roots, however, we find that the gap actually *diminishes* steadily! These gaps are shown below:

0.414, 0.318, 0.268, 0.236, 0.213,
0.196, 0.183, 0.172, 0.162, 0.154,
0.148, 0.141, 0.136, 0.131, 0.127,
0.123, 0.119, 0.116, 0.113, ...

(The 1st entry shown is $\sqrt{2} - 1$ and the last one is $\sqrt{20} - \sqrt{19}$.) Observe how the terms steadily diminish. Further computation only serves to strengthen this observation:

$$\sqrt{101} - \sqrt{100} \approx 0.0499, \qquad \sqrt{1001} - \sqrt{1000} \approx 0.0158.$$

We now show that this pattern is a genuine one; indeed, for all positive integers n,

$$\sqrt{n+1} - \sqrt{n} < \sqrt{n} - \sqrt{n-1}.$$

We shall further show that

$$\sqrt{n+1}-\sqrt{n}<\frac{1}{2\sqrt{n}}.$$

The first inequality may be written in an equivalent form as

$$\sqrt{n+1}+\sqrt{n-1}<2\sqrt{n}.$$

In this inequality, all the quantities involved are positive, so if we square both sides, the inequality will remain unchanged; the quantity on the left side will continue to be smaller than that on the right side. (Note: This is an important point! If there are negative quantities involved in an inequality, then we cannot indiscriminately resort to squaring. For instance, it is true that $-3 < 2$, but if we square both sides of this inequality we get $9 < 4$, which is false. However, when all the quantities involved are positive, then squaring is a valid step, and a reversible one.) So if we prove the modified inequality, then our job is done; the original inequality too will be proved. So we shall try to show that

$$(n+1)+(n-1)+2\sqrt{n^2-1}<4n, \quad \text{or} \quad \sqrt{n^2-1}<n.$$

But this is obviously true (to see why, we again resort to squaring: $n^2-1<n^2$, $\therefore \sqrt{n^2-1}<n$), so the original inequality itself must be true: $\sqrt{n+1}-\sqrt{n}<\sqrt{n}-\sqrt{n-1}$. Therefore, the gap between the square roots of consecutive integers does diminish steadily, as the numerical evidence had suggested.

The second inequality enables us to place upper bounds on the gaps. We may chance upon the result by examining the data. Thus, the figures

$$\begin{aligned}
\sqrt{10}-\sqrt{9} &\approx 0.16,\\
\sqrt{40}-\sqrt{39} &\approx 0.08,\\
\sqrt{160}-\sqrt{159} &\approx 0.04,\\
\sqrt{640}-\sqrt{639} &\approx 0.02,\\
\sqrt{2560}-\sqrt{2559} &\approx 0.01,
\end{aligned}$$

suggest quite obviously that when n grows by a factor of 4, $\sqrt{n}-\sqrt{n-1}$ *diminishes* by a factor of 2 (approximately).

To prove the inequality, we resort once again to squaring; thus to show that

$$\sqrt{n+1}-\sqrt{n}<\frac{1}{2\sqrt{n}}$$

is equivalent to showing that

$$n+1 < \left(\sqrt{n} + \frac{1}{2\sqrt{n}}\right)^2.$$

Simplifying the expression on the right, the inequality reads

$$n+1 < n+1+\frac{1}{4n},$$

which is true. Therefore, the original inequality is correct.

★ ★ ★

There is a more direct way of proving the inequality. Here, we use the fact that

$$\left(\sqrt{n+1}-\sqrt{n}\right) \times \left(\sqrt{n+1}+\sqrt{n}\right) = (n+1)-n = 1.$$

(This draws on the formula $(a-b)\times(a+b) = a^2-b^2$.) The identity may be expressed in another way:

$$\sqrt{n+1}-\sqrt{n} = \frac{1}{\sqrt{n+1}+\sqrt{n}}.$$

Since $\sqrt{n+1} > \sqrt{n}$, we have $\sqrt{n+1}+\sqrt{n} > 2\sqrt{n}$, and so

$$\frac{1}{\sqrt{n+1}+\sqrt{n}} < \frac{1}{2\sqrt{n}}.$$

Combining the two results, we get the inequality we want.

This method gives us more than we asked for—a bonus! It tells us that

$$\frac{1}{2\sqrt{n+1}} < \sqrt{n+1}-\sqrt{n} < \frac{1}{2\sqrt{n}}.$$

So the quantity $\sqrt{n+1}-\sqrt{n}$ exceeds $1/2\sqrt{n+1}$, yet is less than $1/2\sqrt{n}$. We have succeeded in bounding $\sqrt{n+1}-\sqrt{n}$ within a rather narrow interval.

★ ★ ★

If we assume $n = k^2$, then the inequality takes on a more useful form, namely that $\sqrt{k^2+1}-k$ lies between

$$\frac{1}{2\sqrt{k^2+1}} \quad \text{and} \quad \frac{1}{2k}.$$

Noting how close the quantities $\sqrt{k^2+1}$ and k are when k is large, we may use the approximation sign and write:

$$\sqrt{k^2+1}-k \approx \frac{1}{2k}, \quad \text{or} \quad \sqrt{k^2+1} \approx k + \frac{1}{2k}.$$

The approximation gets better as k grows. For instance, when $k = 100$, we have:

$$\sqrt{10001} \approx 100.0049999, \quad 100 + \frac{1}{2 \times 100} = 100.005,$$

and the difference between $\sqrt{10001}$ and $100+1/200$ is approximately 10^{-5} %.

There is yet another way of expressing this result, which proves in the long run to be much more useful. Dividing both sides of the relation $\sqrt{k^2+1} \approx k + 1/2k$ by k, we get

$$\sqrt{1+\frac{1}{k^2}} \approx 1 + \frac{1}{2k^2}.$$

Noting the occurrence of k^2 on both sides, we put $x = 1/k^2$; then when k is large, x is small. So we obtain the following result:

$$\text{when } x \approx 0, \quad \sqrt{1+x} \approx 1 + \frac{x}{2}.$$

To check the closeness of the approximation, put $x = 0.002$; we get $\sqrt{1.002} = 1.0009995$, $1 + 0.002/2 = 1.001$. Close!

⋆ ⋆ ⋆

A few amusing approximations may be extracted from this relation. If we put $x = 1/288$, we get

$$\sqrt{1+x} = \sqrt{1+\frac{1}{288}} = \sqrt{\frac{289}{288}} = \frac{17}{12\sqrt{2}},$$

and

$$1 + \frac{x}{2} = 1 + \frac{1}{2 \times 288} = \frac{577}{576}.$$

This shows that

$$\frac{17}{12\sqrt{2}} \approx \frac{577}{576}, \quad \therefore \sqrt{2} \approx \frac{17 \times 576}{12 \times 577} = \frac{816}{577}.$$

So $816/577$ is a good rational approximation to $\sqrt{2}$. The closeness of the approximation may be seen by displaying the decimal digits of the two numbers:

$$\sqrt{2} = 1.41421356\ldots,$$
$$\frac{816}{577} = 1.41421143\ldots.$$

The two numbers agree till the 5th decimal place.

A better approximation is obtained by taking $x = 9800$. We find, after some manipulations, that

$$\sqrt{2} \approx \frac{27720}{19601}.$$

The gap between $\sqrt{2}$ and 27720/19601 is less than 2×10^{-7} %.

14.2 Sums of square roots

We shall now extract another bonus result from the inequality

$$\frac{1}{2\sqrt{k+1}} < \sqrt{k+1} - \sqrt{k} < \frac{1}{2\sqrt{k}}.$$

Let n be any integer. If we substitute successively $k = n - 1$, $k = n - 2$, $k = n - 3$, …, $k = 2$, $k = 1$ into the inequality, and then add everything together we get:

$$1/(2\sqrt{n}) < \sqrt{n} - \sqrt{n-1} < 1/(2\sqrt{n-1}),$$
$$1/(2\sqrt{n-1}) < \sqrt{n-1} - \sqrt{n-2} < 1/(2\sqrt{n-2}),$$
$$1/(2\sqrt{n-2}) < \sqrt{n-2} - \sqrt{n-3} < 1/(2\sqrt{n-3}),$$

and so on. The last inequality in the list is

$$1/(2\sqrt{2}) < \sqrt{2} - 1 < 1.$$

Now, we add together the corresponding terms of all the inequalities. Nice things happen to the quantity in the middle – everything cancels out, almost! Indeed, all that is left is the quantity $\sqrt{n} - 1$. For convenience, we write

$$a_n = 1 + \frac{1}{\sqrt{2}} + \frac{1}{\sqrt{3}} + \cdots + \frac{1}{\sqrt{n}}.$$

So we have

$$\frac{1}{2}(a_n - 1) < \sqrt{n} - 1 < \frac{1}{2}\left(a_n - \frac{1}{\sqrt{n}}\right).$$

This tells us that a_n is *less* than $2\sqrt{n}-1$ but *more* than $2(\sqrt{n}-1)+1/\sqrt{n}$. So we have a neat result: the sum of the quantities 1, $1/\sqrt{2}$, $1/\sqrt{3}$, ..., $1/\sqrt{n-1}$ and $1/\sqrt{n}$ lies between $2(\sqrt{n}-1)+1/\sqrt{n}$ and $2\sqrt{n}-1$.

★ ★ ★

This result may not seem too impressive to the reader, but in fact it *is* a significant result! The following discussion may help to put things in perspective. We have learnt earlier how to find a formula for the sum $1+2+3+\cdots+n$; the sum is $\frac{1}{2}n(n+1)$. We have likewise learnt how to find a formula for the sum $1^2+2^2+3^2+\cdots+n^2$; the sum is $\frac{1}{6}n(n+1)(2n+1)$. We may similarly work out formulas for sums such as $1^3+2^3+3^3+\cdots+n^3$ and $1^4+2^4+3^4+\cdots+n^4$. These formulas are easy to find, and they are *exact*; there is no element of approximation in them. But for the sum

$$1+\frac{1}{\sqrt{2}}+\frac{1}{\sqrt{3}}+\cdots+\frac{1}{\sqrt{n}},$$

no worthwhile exact formula can be found! Or, to put it more precisely, exact formulas may be found, but they involve functions whose values themselves are hard to get. In such cases, exact formulas are not of much use; approximate formulas are worth a lot more than exact formulas. So our finding has some significance, namely that the sum in question is approximately $2\sqrt{n}-1$. With some hard work, still better formulas may be found.

★ ★ ★

It will be of interest to see if we can produce a similar formula for the quantity

$$s_n = 1+\sqrt{2}+\sqrt{3}+\cdots+\sqrt{n}.$$

The comment made above applies here too: no worthwhile exact formula may be found for the sum.

As a rough first guess, we check whether s_n is approximately equal to the product

number of terms $\times$ the middle term.

Since the number of terms is n and the middle term is roughly $\frac{1}{2}\sqrt{n}$, this is the same as asking whether s_n is approximately equal to $\frac{1}{2}n^{1.5}$. A simple way to check this is to compute the values taken by $s_n/n^{1.5}$ for large values of n. If the values are close to $\frac{1}{2}$, then we may safely conclude that s_n is indeed well-approximated by the formula $\frac{1}{2}n^{1.5}$. A few representative values of s_n and $s_n/n^{1.5}$ are given below:

$$\begin{aligned} s_5 &\approx 8.38, & s_5/5^{1.5} &\approx 0.75, \\ s_{10} &\approx 22.47, & s_{10}/10^{1.5} &\approx 0.71, \\ s_{15} &\approx 40.47, & s_{15}/15^{1.5} &\approx 0.70. \end{aligned}$$

These values certainly do not seem to be close to $\frac{1}{2}$!

But all is not lost—the values do seem to be close to *some* number. And this is a useful finding: if we know that for large n, the value of $s_n/n^{1.5}$ stays close to some number c, then we can safely conclude that $s_n \approx cn^{1.5}$. With this in mind, we compute the values taken by $s_n/n^{1.5}$ when n is large. Table B.14.1 gives some relevant values, for $n = 100, 200, 300, \ldots, 1000$.

Table B.14.1 *Data on sums of square roots*

n	s_n	$n^{1.5}$	$s_n/n^{1.5}$
100	671.46	1000.	0.671
200	1892.48	2828.4	0.669
300	3472.56	5196.2	0.668
400	5343.13	8000.0	0.668
500	7464.53	11180.3	0.668
600	9810.00	14696.9	0.667
700	12359.86	18520.3	0.667
800	15098.88	22627.4	0.667
900	18014.79	27000.0	0.667
1000	21097.46	31622.8	0.667

We make a discovery almost at once: *for large n, the ratio $s_n/n^{1.5}$ stays nearly constant, at about* 0.667. Now, 0.667 is about 2/3, so we formulate our finding thus:

For large n, $s_n/n^{1.5}$ is nearly equal to 2/3.

Further computation supports this statement; e.g.,

$$s_{10000} = 666716.46, \qquad s_{10000}/10000^{1.5} = 0.666716.$$

The closeness of the ratio to 2/3 is striking.

The formula $s_n \approx 0.667n^{1.5}$ is certainly a nice discovery, but can it be improved upon? The reason we ask this is that though in terms of percentage difference the quantities

$$s_n, \quad \frac{2}{3}n^{1.5}$$

are close, in terms of *absolute difference* the two are not too close to one another. Thus, consider $n = 10000$; we have

$$s_{10000} = 666716.46, \qquad \frac{2 \times 10000^{1.5}}{3} = 666666.7.$$

The difference between the two quantities is approximately 49.8; this is not insignificant! Can we do better? Indeed we can, and an extremely detailed analysis shows that an excellent formula is

$$s_n \approx \sqrt{n+1}\left(\frac{2n}{3} + \frac{1}{6} + \frac{1}{24(n+1)}\right).$$

A remarkable formula! Let us check what the formula reveals for $n = 10000$. We already know that $s_{10000} = 666716.46$. The quantity on the right takes the value

$$\begin{aligned} &\sqrt{10001}\left(\frac{20000}{3} + \frac{1}{6} + \frac{1}{24 \times 10001}\right) \\ = \; &100.0049999 \times 6666.833337 \; = \; 666716.67. \end{aligned}$$

Very close! – the difference is roughly 0.21.

Curiously, the formula yields good results even for small values of n. Thus, let $n = 10$; we get $s_{10} \approx 22.4683$, and

$$\sqrt{11}\left(\frac{20}{3} + \frac{1}{6} + \frac{1}{24 \times 11}\right) = 3.31662 \times 6.83712 = 22.6762.$$

The error is again roughly 0.21; more curious! An even better formula than the above is

$$s_n \approx \sqrt{n+1}\left(\frac{2n}{3} + \frac{1}{6} + \frac{1}{24(n+1)}\right) - 0.208.$$

The reader may check the accuracy of this formula using a computer.

Unfortunately, the analysis behind this formula is rather too involved to present here. We require an advanced branch of mathematics known as *calculus*, which provides us with ways of finding the maximum and minimum values of functions, for finding the areas enclosed by curved regions, for computing the lengths of curves, for computing the volumes of curved bodies, and for finding the behaviour of sums such as the one studied above.

⋆⋆⋆

We now present an arithmetical procedure that establishes links with the sequence of square roots in an extremely curious manner. Let $u_1, u_2, u_3, \ldots$ be a sequence of numbers computed according to the following rule: $u_1 = 1$, and

$$u_n = u_{n-1} + \frac{1}{u_{n-1}} \quad \text{for } n = 2, 3, 4, \ldots.$$

So $u_2 = 2$, $u_3 = 2.5$, $u_4 = 2.9$, and so on. Using a computer, we find more terms of the u-sequence. Here are a few representative values, for $n = 100, 200, 300, \ldots, 1000$.

n	u_n
100	14.21371
200	20.05928
300	24.54744
400	28.33231
500	31.66751

n	u_n
600	34.68317
700	37.45665
800	40.03833
900	42.46324
1000	44.75689

Asking for an exact formula for u_n turns out to be none too worthwhile a task. Instead, we shall search for an approximate formula, in the spirit of the discussion conducted above. Our first result is: u_n *is approximately equal to* $\sqrt{2n}$. In support of this surprising statement, we have the following figures: (a) let $n = 100$; we have $u_{100} \approx 14.2137$ and $\sqrt{2 \times 100} \approx 14.1421$; (b) let $n = 500$; we have $u_{500} \approx 31.6675$ and $\sqrt{2 \times 500} \approx 31.6228$. Quite close!

To prove the result $u_n \approx \sqrt{2n}$, we use the principle of induction. A first and rather crude way is the following. Suppose that for some reasonably large value of k we have $u_k \approx \sqrt{2k}$. Then, using the definition, $u_{k+1} = u_k + 1/u_k$,

$$u_{k+1}^2 = u_k^2 + 2 + \frac{1}{u_k^2} \approx 2k + 2 + \frac{1}{2k}.$$

Since the quantity $1/2k$ is small, we ignore it and write $u_{k+1}^2 \approx 2k+2$. So if $u_k \approx \sqrt{2k}$, then $u_{k+1} \approx \sqrt{2(k+1)}$. The approximation carries forward to the next value of k and allows us to conclude, inductively, that $u_n \approx \sqrt{2n}$ for all n.

While this argument does give us a handle on the analysis, it is much too crude. We have not kept track of the errors, and the errors may start to accumulate as we go further down the sequence, thereby spoiling the approximation; so a finer analysis is required. Fortunately, the idea used above may be used again. We first establish the following using induction: $u_n > \sqrt{2n}$ *for all* $n \geq 3$.

Observe that we have written $n \geq 3$; indeed, the inequality is false for $n = 1$ and 2. The proof is simple. We have $u_3 = 2.5 > \sqrt{6}$ $(= 2.44949)$. Next, if $u_n > \sqrt{2n}$, then

$$u_{n+1}^2 = u_n^2 + 2 + \frac{1}{u_n^2} > 2n + 2,$$

since $1/u_n^2$ is a positive quantity. So $u_{n+1} > \sqrt{2(n+1)}$. This establishes the inductive step and so the claim is proved.

Combining the relations $u_{n+1}^2 = u_n^2 + 2 + 1/u_n^2$ and $u_n \approx \sqrt{2n}$, we obtain $u_{n+1}^2 - u_n^2 \approx 2 + 1/2n$. So we have, in succession,

$$\begin{array}{lcl} u_n^2 - u_{n-1}^2 & \approx & 2 + 1/2(n-1) \\ u_{n-1}^2 - u_{n-2}^2 & \approx & 2 + 1/2(n-2) \\ \cdots\cdots\cdots & \cdots & \cdots\cdots\cdots \\ u_2^2 - u_1^2 & \approx & 2 + 1/2. \end{array}$$

The approximation worsens as we move down to smaller values of n, but let us ignore this for the moment. Adding together the corresponding sides of all these relations, we get (after much cancellation)

$$u_n^2 - u_1^2 \approx 2(n-1) + \frac{1}{2}\left(1 + \frac{1}{2} + \cdots + \frac{1}{n-1}\right).$$

Since $u_1 = 1$, this yields the curious relation

$$u_n \approx \sqrt{2n - 1 + \tfrac{1}{2}\left(1 + \tfrac{1}{2} + \cdots + \tfrac{1}{n-1}\right)}.$$

This approximation turns out to be substantially better than $u_n \approx \sqrt{2n}$. The table below gives some representative values of u_n and the quantity b_n given by the RHS of the formula shown above.

n	u_n	b_n
500	31.66751	31.66063
1000	44.75689	44.75201
1500	54.80313	54.79913
2000	63.27344	63.26997

The closeness of the approximation is plainly visible. Is it not remarkable that the approximation is so good, especially considering the drastic approximations we made at the start?

14.3 A square root spiral

Here is an attractive visual which we can produce with square roots, using the Pythagorean theorem. On a sheet of paper, we draw a line segment OP_1 of unit length, and then a segment P_1P_2 of unit length, perpendicular to OP_1. The length of OP_2 is given by $OP_2^2 = 1^2 + 1^2$, $\therefore\ OP_2 = \sqrt{2}$.

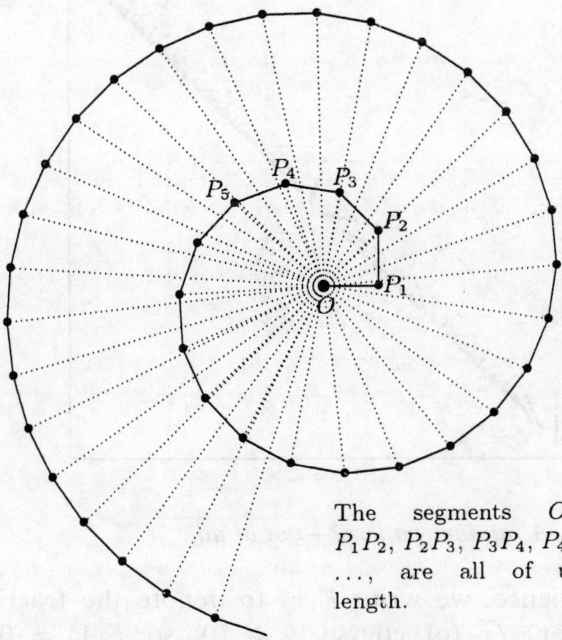

The segments OP_1, P_1P_2, P_2P_3, P_3P_4, P_4P_5, ..., are all of unit length.

Figure B.14.1 *A "square root spiral"*

Next, we draw a segment P_2P_3, once again of unit length, perpendicular to OP_2. The length of OP_3 is given by $OP_3^2 = 1^2 + (\sqrt{2})^2 = 3$, $\therefore OP_3 = \sqrt{3}$.

We continue in this way; having located P_{n-1}, we draw a segment $P_{n-1}P_n$ of unit length, perpendicular to OP_{n-1}. Each time, we turn in the anti-clockwise direction. It is clear that the length of OP_n is $\sqrt{n}$. When we have drawn sufficiently many of the segments P_iP_{i+1}, they are seen to form a pretty and elegant spiral, as shown in Figure B.14.1.

14.4 A random sequence?

We conclude the chapter on square roots with a pretty and quite unexpected picture. Earlier on, we had commented that the sequence of decimal parts of the sequence $\{\sqrt{n}\}$ has a "random" appearance. This is not true in the strict sense of the word, but nevertheless there is an element of truth in it. We explain below what we mean by this.

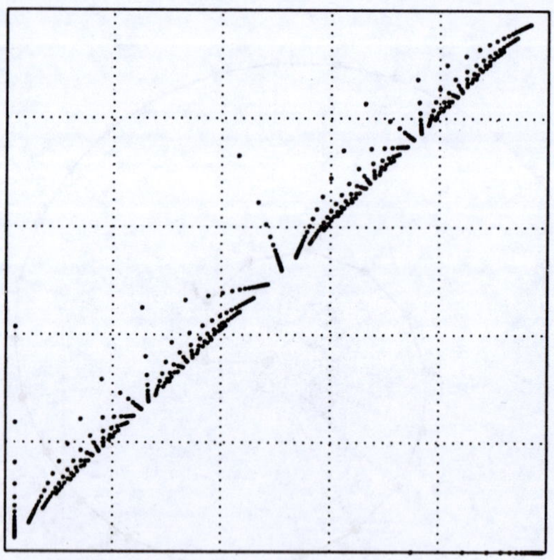

Figure B.14.2 *A random pattern?—not at all!*

For convenience, we write $F(x)$ to denote the fractional part of the number $\sqrt{x}$ (of course, $x \geq 0$); so $F(1) = 0$, $F(2) = \sqrt{2} - 1 = 0.414213\ldots$, $F(3) = \sqrt{3} - 1 = 0.713205\ldots$, $F(4) = 0$, $F(5) = \sqrt{5} - 2 = 0.236068\ldots$, and so on. We now form pairs of numbers as shown below,

$$(F(0), F(1)),\ (F(1), F(2)),\ (F(2), F(3)),\ (F(3), F(4)),\ \ldots,$$

and regard these as *coordinates of points*. What happens when we plot these points on a sheet of graph paper? Since $F(x)$ lies between 0 and 1 for all x, the points all lie within the unit square [whose vertices are the points $(0,0)$, $(1,0)$, $(1,1)$ and $(0,1)$). Will the points be scattered randomly within the square? When we perform the exercise of plotting the first 400 such points, the answer comes as a major surprise. (See Figure B.14.2.)

How is one to account for this very curious pattern? The streaks running in various directions from the centre of the square are particularly difficult to explain. We shall leave to the the reader the pleasant task of finding a coherent explanation behind this phenomenon.

However, a slight variation produces a figure which looks "random". Instead of plotting the points $(F(n), F(n+1))$ for $n = 0$, 1, 2, ..., 400, we plot the points $(F(n), F(2n))$ for the same range of values of n. That is, we plot the points

$$(F(0), F(0)),\ (F(1), F(2)),\ (F(3), F(6)),\ (F(4), F(8)),\ \ldots.$$

The result is shown below (Figure B.14.3).

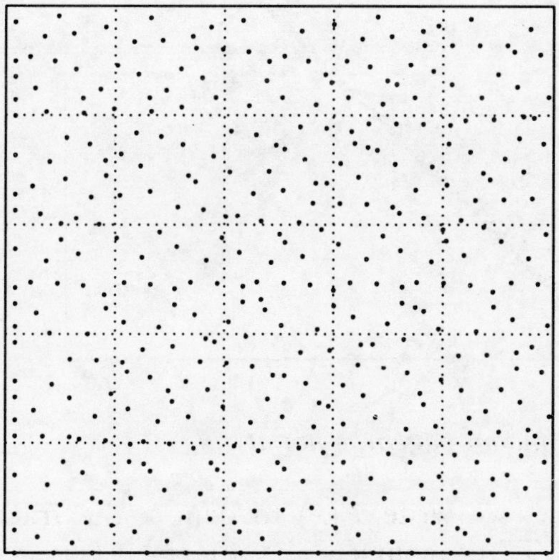

Figure B.14.3 *A random sequence?—yes!*

Now, it does indeed look like dust scattered randomly over the unit square! (But here too one sees streaks, running in a north-easterly direction; they are difficult to explain.)

Exercises

14.4.1 Show that $\sqrt{3}$ is an irrational number.

14.4.2 Show that $\sqrt{6}$ is irrational.

14.4.3 Show that the cube root of 2 is irrational.

Chapter 15

Fibonacci Numbers

15.1 Family planning

The following problem was first posed in the book *Liber Abaci* written in 1225 by Leonardo of Pisa (1180–1250), also known as Fibonacci ("son of Bonaccio").

> *How many pairs of rabbits will be produced in a year, beginning with a single pair, if in every month each pair bears a new pair which itself becomes productive from the second month onwards?*

Instead of asking only for the population at the end of a year, we shall ask, more generally, how many pairs are there at the end of n months? The answer is to be given in terms of n. (Note that the rabbits are assumed to be immortal!)

Let the total number of pairs of rabbits at the start of the nth month be denoted by the symbol Fib_n. This number is the nth *Fibonacci number*, named in honour of Fibonacci. To get a feel for these numbers, we shall compute the first few by hand.

Of course, $\mathrm{Fib}_1 = 1$ (this is the starting pair) and $\mathrm{Fib}_2 = 1$. (The population explosion has not begun yet!) The original pair gives birth to a new pair at the start of month 3, so $\mathrm{Fib}_3 = 2$. At the start of month 4, the original pair has given birth to one more pair and no other pair has begotten any offspring, so $\mathrm{Fib}_4 = 3$. At the start of month 5, the original pair has given birth yet again to a pair, and so has the pair which was born at the end of the second month, so $\mathrm{Fib}_5 = 5$. Continuing these calculations, we obtain the following:

$$\begin{array}{llll} \mathrm{Fib}_1 = 1, & \mathrm{Fib}_2 = 1, & \mathrm{Fib}_3 = 2, & \mathrm{Fib}_4 = 3, \\ \mathrm{Fib}_5 = 5, & \mathrm{Fib}_6 = 8, & \mathrm{Fib}_7 = 13, & \mathrm{Fib}_8 = 21, \\ \mathrm{Fib}_9 = 34, & \ldots & \ldots & \ldots . \end{array}$$

The pattern is unmistakable. We shall refer to this sequence of numbers as the *Fibonacci sequence.*

⋆ ⋆ ⋆

The defining relation To compute more numbers of the sequence we must find a more efficient way than that used above; we need to find a recurrence relation for the sequence.

Consider any two consecutive months, say month n and month $(n+1)$; at the start of month n there are Fib_n rabbit pairs, and at the start of month $(n+1)$ there are Fib_{n+1} pairs. So the number of rabbit pairs born during month n is $\text{Fib}_{n+1} - \text{Fib}_n$. Now, who are the parents of these rabbits?—clearly the rabbits who were present *two* months prior to that month-end; in other words, the rabbit pairs who were present at the end of month $(n-2)$; that is, at the start of month $(n-1)$. By definition, the number of these rabbit pairs is Fib_{n-1}, so it follows that $\text{Fib}_{n+1} - \text{Fib}_n = \text{Fib}_{n-1}$; in other words,

$$\text{Fib}_{n+1} = \text{Fib}_n + \text{Fib}_{n-1}.$$

This is therefore the defining relation for the Fibonacci sequence. So the next few Fibonacci numbers, after Fib_9, are:

$$\text{Fib}_{10} = 55, \quad \text{Fib}_{11} = 89, \quad \text{Fib}_{12} = 144, \quad \ldots.$$

The table below displays the first forty Fibonacci numbers.

1,	1,	2,	3,	5,
8,	13,	21,	34,	55,
89,	144,	233,	377,	610,
987,	1597,	2584,	4181,	6765,
10946,	17711,	28657,	46368,	75025,
121393,	196418,	317811,	514229,	832040,
1346269,	2178309,	3524578,	5702887,	9227465,
14930352,	24157817,	39088169,	63245986,	102334155.

Observe the impressive growth rate of the sequence—the 40th term is greater than one hundred million!

15.2 A formula for Fib_n

In this section, we shall derive a formula for the nth Fibonacci number. The derivation will be suggestive rather than rigorous, and we shall proceed in a somewhat roundabout manner.

Consider the number $r = 2$. Observe that r satisfies the relation $r^2 = r + 2$. Multiplying both sides of this relation by r^{n-2}, we see that

$$r^n = r^{n-1} + 2r^{n-2}, \qquad \text{for all } n \geq 2.$$

This means that for the sequence of powers of 2, namely 1, 2, 4, 8, 16, 32, 64, ..., if we denote the nth term by a_n, then we have the following relation for a_n:

$$a_n = a_{n-1} + 2a_{n-2}, \qquad \text{for all } n \geq 2.$$

For example, $32 = 16 + (2 \times 8)$, $64 = 32 + (2 \times 16)$, and so on.

Now, there is another number r which satisfies the equation $r^2 = r + 2$; namely, $r = -1$. This means that the sequence of powers of -1 satisfies the very same recursive relation! That is, for the sequence $-1, 1, -1, 1, -1, 1, -1, 1, \ldots$, denoting the nth term by b_n, we have

$$b_n = b_{n-1} + 2b_{n-2}, \qquad \text{for all } n \geq 2.$$

Thus, $1 = -1 + (2 \times 1)$ and $-1 = 1 + (2 \times -1)$.

There is still more that can be said: *any sequence that can be generated via addition from these two sequences (the powers of 2 and the powers of -1) satisfies the very same recursion.* For instance, consider the sequence whose nth term a_n is

$$a_n = (3 \times 2^n) + (5 \times (-1)^n).$$

The first few a-values are $a_0 = 8$, $a_1 = 1$, $a_2 = 17$, $a_3 = 19$, $a_4 = 53$, ..., so the sequence is: 8, 1, 17, 19, 53, 91, 197, ... Observe that $a_n = a_{n-1} + 2a_{n-2}$.

We can make more such statements, using different values of r. For instance if $r = 3$, then we have $r^2 = 2r + 3$, so

$$r^n = 2r^{n-1} + 3r^{n-2}, \qquad \text{for all } n \geq 2.$$

Thus, for the sequence of the powers of 3, namely 1, 3, 9, 27, 81, 243, ..., if we denote the nth term by c_n, then we have the following recursive relation for c_n:

$$c_n = 2c_{n-1} + 3c_{n-2}, \qquad \text{for all } n \geq 2.$$

For example, $27 = (2 \times 9) + (3 \times 3)$, $81 = (2 \times 27) + (3 \times 9)$, and so on. And since the equation $r^2 = 2r + 3$ also has the solution

$r = -1$, the sequence of powers of -1 satisfies this very same recursive relationship.

Of course, the number $r = 3$ satisfies infinitely many such equations; e.g., the equation $r^2 = 4r - 3$. This means that the sequence $\{c_n\}$ also satisfies the following recursion:

$$c_n = 4c_{n-1} - 3c_{n-2}, \qquad \text{for all } n \geq 2.$$

For example, $81 = (4 \times 27) - (3 \times 9)$, and so on.

★ ★ ★

Now the recursive relationship for the Fibonacci numbers is $\text{Fib}_n = \text{Fib}_{n-1} + \text{Fib}_{n-2}$. It would seem therefore that the relevant equation to be solved is $r^2 = r + 1$. This equation, unfortunately, does not have integer solutions. In fact, the solutions of this equation are the numbers

$$\frac{1+\sqrt{5}}{2} = 1.6180339\ldots, \qquad \frac{1-\sqrt{5}}{2} = -0.6180339\ldots.$$

We shall denote the number $(1+\sqrt{5})/2$ by ϕ; thus, $\phi \approx 1.6180339$. *This is the so-called "golden ratio".* Observe that in terms of ϕ, the other solution of the equation may be written as $-1/\phi$. This may be checked by verifying that

$$\frac{1+\sqrt{5}}{2} \cdot \frac{1-\sqrt{5}}{2} = -1.$$

The number ϕ has innumerable nice properties that serve to justify its name! For instance, we have

$$\phi^2 = \phi + 1 = 2.6180339\ldots, \quad \frac{1}{\phi} = \phi - 1 = 0.6180339\ldots,$$

and we also have:

$$\phi^3 = 2\phi + 1,\ \phi^4 = 3\phi + 2,\ \phi^5 = 5\phi + 3,\ \phi^6 = 8\phi + 5,$$

and so on.

It should not come as a surprise that for the sequence 1, ϕ, ϕ^2, ϕ^3, ϕ^4, ..., if we denote the nth term by a_n, then we have $a_n = a_{n-1} + a_{n-2}$ for $n \geq 2$. Moreover, the very same statement can be made for the powers of the other solution of the equation

$r^2 = r + 1$; that is, for the powers of $-1/\phi$. Therefore, *for any two numbers c and d, the sequence* $\{a_n\}$ *for which*

$$a_n = c \cdot \phi^n + d \cdot \left(\frac{-1}{\phi}\right)^n \qquad (n = 1, 2, 3, \ldots),$$

satisfies the recursion $a_n = a_{n-1} + a_{n-2}$.

It follows from this that if we can find two numbers c and d, such that a_1 and a_2 coincide with the first two Fibonacci numbers (Fib_1 and Fib_2), then the equation

$$\text{Fib}_n = c \cdot \phi^n + d \cdot \left(\frac{-1}{\phi}\right)^n$$

will hold for all positive integers n. That is, we would have found a generating formula for the Fibonacci sequence. This holds because if the 1st and 2nd terms of the a-sequence are the same as the 1st and 2nd terms of the Fibonacci sequence, and the recursive relation is the same, then the two sequences will agree in the 3rd term also, therefore in the 4th term too, therefore in the 5th term, and so on, indefinitely.

But this task is easily performed; all we need to do is to find the constants c and d, such that

$$c \cdot \phi + d \cdot \left(\frac{-1}{\phi}\right) = 1, \qquad c \cdot \phi^2 + d \cdot \left(\frac{1}{\phi^2}\right) = 1.$$

Since $1/\phi = \phi - 1$ and $1/\phi^2 = (\phi - 1)^2 = -\phi + 2$, the equations may be rewritten more conveniently as

$$\begin{aligned} c\phi - d(\phi - 1) &= 1, \\ c(\phi + 1) + d(\phi + 2) &= 1. \end{aligned}$$

These equations are easy to solve, we get

$$c = \frac{1}{2\phi - 1}, \quad d = \frac{1}{1 - 2\phi} = -c.$$

Substituting for $\phi = (1 + \sqrt{5})/2$, we get $c = 1/\sqrt{5}$ and $d = -1/\sqrt{5}$. It follows that the nth Fibonacci number is given by the formula

$$\text{Fib}_n = \frac{1}{\sqrt{5}}\phi^n - \frac{1}{\sqrt{5}}\left(\frac{-1}{\phi}\right)^n.$$

This is known as *Binet's formula*. It is certainly quite impressive! But we can cast it in a simpler and more convenient form by arguing as follows. Note that the quantity $(1/\sqrt{5}) \times 1/\phi$ has the approximate value 0.276. This is a small number and, in particular, it is less than 0.5. Therefore,

$$\text{Fib}_n = \frac{\phi^n}{\sqrt{5}} \pm \text{ some number less than } 0.5.$$

Since Fib_n is a whole number, the following conclusion becomes inevitable:

$$\text{Fib}_n = \text{the integer closest to } \frac{\phi^n}{\sqrt{5}}.$$

The formula is easy to verify. For instance, let $n = 10$; then

$$\frac{\phi^{10}}{\sqrt{5}} \approx 55.0036,$$

and the integer closest to 55.0036 is 55, which is indeed what Fib_{10} equals. And for $n = 13$, we get

$$\frac{\phi^{13}}{\sqrt{5}} \approx 232.999,$$

and the integer closest to 232.999 is 233, which is the value of Fib_{13}. This is indeed a triumph for us!

The approach we have used here is a perfectly general one, and may be used for any such sequence.

Remark There are other formulas for the Fibonacci sequence; see Section 15.5.

15.3 Properties of the Fib numbers

The Fibonacci sequence has a vast number of properties, and we could easily fill up a book with a catalogue of its features. A few of the better-known ones are listed here.

1. Let a, b, c be any three consecutive Fibonacci numbers; then $b^2 - ac = \pm 1$.

   Example Consider the consecutive Fibonacci numbers $3, 5, 8$ and $13, 21, 34$. We have, $5^2 - (3 \times 8) = 1$ and $21^2 - (13 \times 34) = -1$.

2. Let a, b, c, d be any four consecutive Fibonacci numbers; then $ad - bc = \pm 1$.

   Example Consider the consecutive Fibonacci numbers 2, 3, 5, 8 and 21, 34, 55, 89. We have $(2 \times 8) - (3 \times 5) = 1$ and $(21 \times 89) - (34 \times 55) = -1$.

3. For all $n \geq 1$ we have the following relation:

$$\mathrm{Fib}_1 + \mathrm{Fib}_3 + \mathrm{Fib}_5 + \cdots + \mathrm{Fib}_{2n-1} = \mathrm{Fib}_{2n}.$$

   Example $\mathrm{Fib}_1 + \mathrm{Fib}_3 + \mathrm{Fib}_5 = 1 + 2 + 5 = 8 = \mathrm{Fib}_6$.

4. For all $n \geq 1$,

$$\mathrm{Fib}_1^2 + \mathrm{Fib}_2^2 + \mathrm{Fib}_3^2 + \cdots + \mathrm{Fib}_n^2 = \mathrm{Fib}_n \mathrm{Fib}_{n+1}.$$

   Example Let $n = 5$; the sum on the left side is

$$1^2 + 1^2 + 2^2 + 3^2 + 5^2 = 40,$$

   and 40 equals $\mathrm{Fib}_5 \times \mathrm{Fib}_8$ (that is 5×8).

5. Every positive integer can be written as a sum of distinct Fibonacci numbers.

   Example The number 100 may be expressed as $100 = 89 + 8 + 3$, while 200 may be expressed as $200 = 144 + 55 + 1$.

6. Consecutive Fibonacci numbers are coprime.

   Example 21 and 34 are coprime (even though neither number is prime), and so are 34 and 55.

7. If m is a divisor of n, then Fib_m is a divisor of Fib_n.

   Example Let $m = 5$ and $n = 15$; then $m \mid n$, $\mathrm{Fib}_m = 5$, $\mathrm{Fib}_n = 610$, and of course we have $5 \mid 610$.

   Or let $m = 7$ and $n = 21$; then $m \mid n$, $\mathrm{Fib}_m = 13$ and $\mathrm{Fib}_n = 10946$, and certainly we have $13 \mid 10946$.

8. If Fib_n is a prime number, then n itself is prime (but the converse is not true).

   Example $\mathrm{Fib}_7 = 13$ is prime, and so is 7; $\mathrm{Fib}_{11} = 89$ is prime, and so is 11.

However, Fib_{19} and Fib_{31} are composite, even though 19 and 31 are prime:

$$\mathrm{Fib}_{19} = 4181 = 37 \times 113,$$
$$\mathrm{Fib}_{31} = 1346269 = 557 \times 2417.$$

A computer search reveals that the primes p below 200 for which Fib_p is prime are the following:

3, 5, 7, 11, 13, 17, 23,

29, 43, 47, 83, 131, 137.

Remark It is not known whether or not there are infinitely many Fibonacci numbers which are prime.

9. If the gcd of m and n is k, then the gcd of Fib_m and Fib_n is Fib_k.

   Example Let $m = 15$ and $n = 20$; then $k = 5$. We have $\mathrm{Fib}_m = 610$ and $\mathrm{Fib}_n = 6765$, and $\mathrm{Fib}_k = 5$; and the gcd of 610 and 6765 is indeed 5. (Proof: 610 and 6765 are divisible by 5, and on division by 5 we get the numbers 122 and 1353. These two numbers are coprime, for $122 = 2 \times 61$ and $1353 = 3 \times 11 \times 41$.)

10. For $n \geq 4$, the number $\mathrm{Fib}_n + 1$ is always composite.

    Example The numbers $\mathrm{Fib}_n + 1$ for $n \geq 4$ are the following:

    6, 9, 14, 22, 35, 56, 90, 145, 234, 378, 611, ...

    Observe that these numbers are all composite. (The number 611 has the factorization 13×47.)

Finally, here are two curiosities (which are rather difficult to prove): *Leaving aside the initial '1', the only square number in the Fibonacci sequence is 144, and the only cube is 8.*

15.4 The golden ratio

Considering the fact that Fib_n is "close" to $\phi^n/\sqrt{5}$, the following should come as no surprise:

$$\frac{\mathrm{Fib}_{n+1}}{\mathrm{Fib}_n} \approx \phi.$$

In fact this is true, but nevertheless there are some pleasant surprises in store for us. Write, for convenience, u_n for the fraction $\mathrm{Fib}_{n+1}/\mathrm{Fib}_n$. Our calculations yield the following: $u_1 = 1$, $u_2 = 2$, $u_3 = 1.5$, $u_4 \approx 1.667$, and so on. The u-sequence is displayed below:

$$1,\ 2,\ 1.5,\ 1.667,\ 1.6,\ 1.615,\ 1.619,\ 1.6176,\ 1.6182,\ \ldots .$$

The sequence visibly oscillates about the value of ϕ and slowing down, zeroes in on ϕ.

Observe that the successive u-values are related by the simple formula

$$u_{n+1} = 1 + \frac{1}{u_n}.$$

Based on this observation, we may derive an amusing yet most revealing expression for ϕ. Observe that $u_1 = 1$ and $u_2 = 1+1/u_1 = 1+1/1$. Next,

$$u_3 = \frac{3}{2} = 1 + \frac{1}{u_2} = 1 + \cfrac{1}{1 + \cfrac{1}{1}}.$$

Continuing, we have

$$u_4 = \frac{5}{3} = 1 + \frac{1}{u_3} = 1 + \cfrac{1}{1 + \cfrac{1}{1 + \cfrac{1}{1}}},$$

and

$$u_5 = \frac{8}{5} = 1 + \frac{1}{u_4} = 1 + \cfrac{1}{1 + \cfrac{1}{1 + \cfrac{1}{1 + \cfrac{1}{1}}}}.$$

We may proceed indefinitely in this manner. Fractions of this type, where the numerators are all 1 and the denominators continue "downwards", are called *simple continued fractions*, and they form an interesting subject of study in algebra and number theory. Readers wishing to learn more about continued fractions should consult the book by Barnard and Child, or the one on number theory by Hardy and Wright. (See the reference section for details.)

The continued fractions for u_4 and u_5 may be written in a more convenient form as follows:

$$u_4 = 1 + \frac{1}{1+}\,\frac{1}{1+}\,\frac{1}{1},$$

$$u_5 = 1 + \frac{1}{1+}\,\frac{1}{1+}\,\frac{1}{1+}\,\frac{1}{1}.$$

The reason for the greater convenience is, of course, purely typographic. It becomes difficult for the typesetter to stack up those 1s in the denominator!

The following result is now forced upon us – the golden number ϕ may be written as

$$\phi = 1 + \frac{1}{1+}\,\frac{1}{1+}\,\frac{1}{1+}\,\frac{1}{1+}\,\frac{1}{1+}\,\frac{1}{1+}\,\frac{1}{1+}\,\frac{1}{1+}\,\frac{1}{1+}\cdots;$$

that is, as an *infinite continued fraction containing only 1s*.

⋆ ⋆ ⋆

Curiously, there is another, quite different, way of expressing ϕ using infinitely many 1s. Recall that ϕ is the solution of the equation $r^2 = r + 1$. This equation may be rewritten as $r = \sqrt{1+r}$. So, we have,

$$\phi = \sqrt{1+\phi}.$$

Note that we have found a *recursive definition for* ϕ; that is, we have expressed ϕ in terms of itself.

We now replace the second occurrence of ϕ in this equation by using this very same equation. We get

$$\phi = \sqrt{1+\sqrt{1+\phi}}.$$

This may, of course, be continued! So, we also have

$$\phi = \sqrt{1+\sqrt{1+\sqrt{1+\phi}}},$$

$$\phi = \sqrt{1+\sqrt{1+\sqrt{1+\sqrt{1+\phi}}}},$$

and so on. In the limit, continuing this infinitely many times, we get

$$\phi = \sqrt{1+\sqrt{1+\sqrt{1+\sqrt{1+\sqrt{1+\cdots}}}}}.$$

What a curious expression!

★ ★ ★

The number ϕ has so many interesting properties that it has also attracted a lot of attention in non-mathematical circles; in other words, from cranks! The suggestion has been made that as it occurs so very widely, it ought to be designated as a universal constant, like π or e.

★ ★ ★

A rectangle whose sides are in the proportion $\phi : 1$ is called a *golden rectangle*. If, from a golden rectangle (of dimensions $1 \times \phi$), a 1×1 square is cut away (this being the largest square that can be cut off from the rectangle), then the rectangle left over is once again golden in shape. To see why this is so, note that the smaller rectangle has the dimensions $(\phi - 1) \times 1$ and, of course, we have the relation

$$(\phi - 1) : 1 = 1 : \phi$$

which follows from the defining equation $\phi^2 = \phi + 1$. (See Figure B.15.1.)

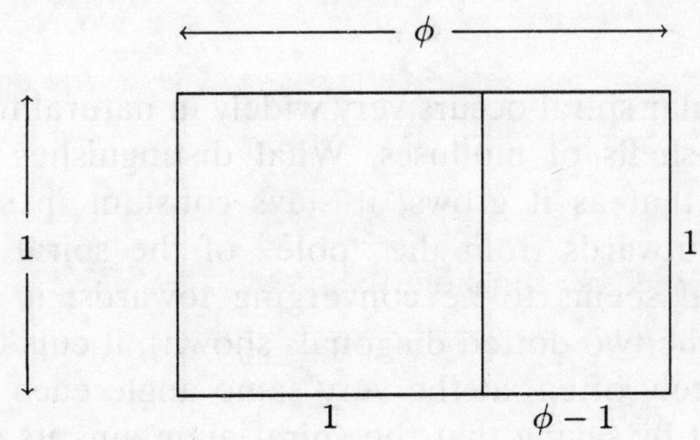

Figure B.15.1 *A golden rectangle*

Naturally, we can repeat this step of cutting off the largest possible square; so we remove the largest possible square from the left-over rectangle, which is itself golden; and then we cut off the largest possible square from the new left-over rectangle; and so on. (We proceed inwards in a clockwise spiral.)

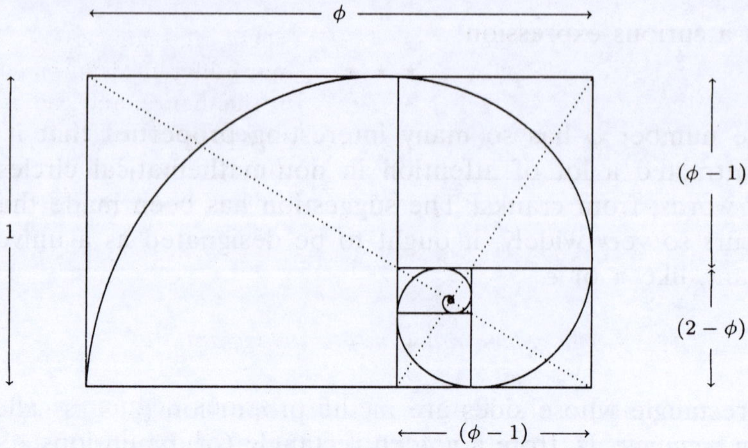

Figure B.15.2 *"Whirling Squares" in the golden rectangle*

The result is shown in the figure above. Martin Gardner refers to the squares in the figure as "whirling squares". It turns out that a spiral may be drawn through some of the vertices of the squares, as shown. This is an example of a *logarithmic spiral*, also known as an *equiangular spiral*.

Remark The figure may give the impression that the spiral touches the sides of each rectangle. In fact this is not the case; there are two points of intersection in each case, but they are rather close to one another.

★ ★ ★

The equiangular spiral occurs very widely in natural forms, most notably in the shells of molluscs. What distinguishes this spiral from others is that as it grows, it stays constant in shape. If a line is drawn outwards from the "pole" of the spiral (the point where the spiral seems to be converging towards; it lies at the intersection of the two dotted diagonals shown), it cuts the coils of the spiral infinitely often, at the very same angle each time. This is what is meant by saying that the spiral maintains its shape as it grows, and it may explain why it is such a favoured form in nature. As the mollusc grows, its proportions are naturally maintained, and an equiangular spiral is the result.

★ ★ ★

If a regular pentagon is drawn, together with all its diagonals, many different line segments may be seen in the diagram.

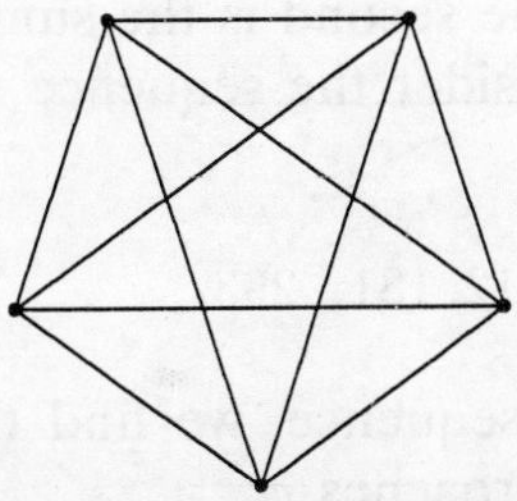

There are segments of four different lengths in this figure, and the ratio of each length to the one just smaller than it is the golden ratio, ϕ.

Figure B.15.3 *The golden ratio and the pentagram*

If we compute the lengths of the various segments, we find, most remarkably, that the ratio of each length to the next smaller length is the golden ratio. This may explain the curious appeal of the pentagram figure, especially its associations with witchcraft. (See Figure B.15.3.)

⋆ ⋆ ⋆

In human affairs, celebrities are often the object of a great deal of (unwelcome) attention; and so also in mathematics. The number ϕ may certainly be regarded as a celebrity, and the most fantastic claims have been made about it. The claim that the most pleasing and most preferred shape of a rectangle is that of the golden rectangle is an example. This appears to have become an accepted fact, and an outcome of this is that architects make use of the golden ratio quite often in the design of buildings. However, the evidence in favour of this claim is inconclusive.

Much more amusing is the claim that for adult women, the ratio of their height to the height of their navels is ϕ! And perhaps the most bizarre claim of all draws on the value of $6\phi^2/5$. We find that

$$\frac{6\phi^2}{5} = 3.14164078649987\ldots,$$

a number which is extremely close to π (whose approximate value is 3.1415926). The claim now is that π is "actually" equal to $6\phi^2/5$, i.e., $\pi = 3.14164\ldots$, and not $3.14159\ldots$, as claimed by some!

⋆ ⋆ ⋆

A pleasant and unexpected discovery about the golden ratio is that ϕ is associated with *any* two-term additive sequence, that is, a sequence in which each term after the second is the sum of the preceding two terms. For instance, consider the sequence starting with 7, 1. Its successive terms are

$$7,\ 1,\ 8,\ 9,\ 17,\ 26,\ 43,\ 69,\ 112, 181,\ 293,\ \ldots,$$

and, as we proceed further down the sequence, we find that the ratio of successive terms gradually approaches ϕ:

$$\frac{69}{43} \approx 1.605, \quad \frac{112}{69} \approx 1.623, \quad \frac{181}{112} \approx 1.616, \quad \ldots.$$

This happens irrespective of which two numbers we place at the start of the sequence.

In an analogy with the Fibonacci sequence, we may define a "tribonacci sequence" as a three-term additive sequence, one in which each term after the third is the sum of the three terms which precede it. So, if we start with the numbers 0, 1, 1, then the sequence we get is: 0, 1, 1, 2, 4, 7, 13, 24, 44, 81, 145, 270, 496, ... Here too, we find that the ratio of successive terms begins to approach some fixed value, namely the number 1.839286754.... Thus,

$$\frac{81}{44} \approx 1.84, \quad \frac{145}{81} \approx 1.79, \quad \frac{270}{145} \approx 1.86, \quad \ldots$$

Just as the golden ratio ϕ is the solution of the equation $r^2 = r+1$, the number that occurs here is the solution of the equation $r^3 = r^2+r+1$. A similar phenomenon occurs with sequences where we compute four-term sums. However, none of these numbers have quite the charismatic nature of the golden ratio.

⋆ ⋆ ⋆

We conclude this section with a description of another pretty and unexpected property possessed by ϕ. Let $\{a_n\}$ and $\{b_n\}$ be sequences of positive integers defined as follows:

$$a_n = [n\phi], \quad b_n = [n\phi^2].$$

Here, the notation $[x]$ refers to the largest integer less than or equal to x. For instance, $[2.75] = 2$, $[\pi] = 3$, $[\sqrt{2}] = 1$, and so on.

The first few terms of the two sequences are:

$a_1 = [\phi] = 1$, $b_1 = [\phi^2] = 2$, $a_2 = [2\phi] = 3$, $b_2 = [2\phi^2] = 5$, $a_3 = [3\phi] = 4$, $b_3 = [3\phi^2] = 7$, ..., $a_{10} = [10\phi] = 16$, $b_{10} = [10\phi^2] = 26$,

.... For brevity, write A for $\{a_n\}$ and B for $\{b_n\}$. The first 20 terms of both A and B are displayed below.

$A = \{1, 3, 4, 6, 8, 9, 11, 12, 14, 16, 17, 19, 21, 22, 24, 25, 27, 29, 30, 32, \ldots\}$,

$B = \{2, 5, 7, 10, 13, 15, 18, 20, 23, 26, 28, 31, 34, 36, 39, 41, 44, 47, 49, 52, \ldots, \}$.

Now, for the property in question: we find that the sequences A and B, considered together, account for *every* positive integer. Moreover, *no integer occurs twice*; that is, A and B have no elements in common whatsoever! In short, A and B constitute a "partition" of the set of positive integers.

15.5 Another formula for Fib_n

We saw earlier that the nth Fibonacci number may be expressed as

$$\text{Fib}_n = \frac{1}{\sqrt{5}}\phi^n - \frac{1}{\sqrt{5}}\left(\frac{-1}{\phi}\right)^n.$$

It turns out there are several formulas that will generate the Fibonacci numbers. Here is one such formula. Let ρ denote either root of the equation $x^2 + 3x + 1 = 0$; then

$$\text{Fib}_n = \frac{1 + \rho + \rho^2 + \cdots + \rho^{n-1}}{(1+\rho)^{n-1}}.$$

So the formula asserts that $\text{Fib}_1 = 1/(1+\rho)^0 = 1$ (true), $\text{Fib}_2 = (1+\rho)/(1+\rho) = 1$ (true), $\text{Fib}_3 = (1+\rho+\rho^2)/(1+\rho)^2$ (to be checked), and so on. Note that the formula may also be expressed as

$$\text{Fib}_n = \frac{1-\rho^n}{1-\rho} \times \frac{1}{(1+\rho)^{n-1}}.$$

Verifying this assertion is easy. Let ρ denote either root of the equation $x^2 + 3x + 1 = 0$, and let $f(n)$ denote the right side of the above equation; then the Fibonacci sequence and the f-sequence agree on their first two terms. Now, let us check whether the f-sequence satisfies the same recursion as the Fibonacci sequence. For this, we need to find out whether $f(n) + f(n+1) = f(n+2)$ for all positive integers n. This is the same (after cancelling the factor $1/(1-\rho)$) as checking whether

$$\frac{1-\rho^n}{(1+\rho)^{n-1}} + \frac{1-\rho^{n+1}}{(1+\rho)^n} = \frac{1-\rho^{n+2}}{(1+\rho)^{n+1}}.$$

Cancelling the common factor $(1+\rho)^{n-1}$ and cross-multiplying reduces the equation to

$$(1+\rho)^2(1-\rho^n)+(1+\rho)(1-\rho^{n+1})=1-\rho^{n+2}.$$

After multiplication and further cancellation, the equation to be checked becomes

$$(1-\rho^n)(1+3\rho+\rho^2)=0,$$

and this is true, as ρ is a root of the equation $x^2+3x+1=0$. So the condition to be checked is verified, which means that the f-sequence obeys the same recursive law as the Fib-sequence. Since their first two terms agree, they agree everywhere. So $\mathrm{Fib}_n = f(n)$ for all n.

Exercises

15.5.1 Show that the solutions of the equation $r^2 = r + 1$ are ϕ and $-1/\phi$.

15.5.2 Let $u_n = \mathrm{Fib}_{n+1}/\mathrm{Fib}_n$. Show that

$$u_{n+1} = 1 + \frac{1}{u_n}.$$

15.5.3 (a) Show that if a, b and c are consecutive Fibonacci numbers, then $b^2 - ac = \pm 1$.

(b) Show that if a, b, c and d are consecutive Fibonacci numbers, then $ad - bc = \pm 1$.

15.5.4 Show that consecutive Fibonacci numbers are coprime (they share no factor greater than 1).

15.5.5 Let the Fibonacci numbers be divided by 2 and only the remainders retained. Show that the remainders repeat after every three terms.

Similarly, if the Fibonacci numbers are divided by 3, then the remainders repeat after every eight terms.

15.5.6 If the successive Fibonacci numbers are divided by 10, then the remainders will be found to repeat after every K terms, for some number K. Find K. (You may need to use a computer.)

15.5.7 Use the method described in the chapter to find a formula for the nth term of the sequence $\{a_n\}$ displayed below:

$$4, 5, 7, 11, 19, 35, 67, 131, 259, \ldots.$$

The recursion relation here is $a_n = 3a_{n-1} - 2a_{n-2}$ for $n \geq 2$.

15.5.8 Use the method described in the chapter to find a formula for the nth term of the sequence $\{b_n\}$ displayed below:

$$1, 3, 7, 17, 41, 99, 239, \ldots.$$

The recursion relation here is $b_n = 2b_{n-1} + b_{n-2}$ for $n \geq 2$.

15.5.9 (a) Show that the ratio of the lengths of the two dotted diagonals in Figure B.15.3 is the golden ratio.

(b) Prove the statement made in the text about the ratios of the lengths of the various segments occurring in a pentagram; namely, that the ratio of each length to the one just smaller than it is ϕ.

15.5.10 *An open-ended investigation*

Study the tribonacci sequence for properties of the type possessed by the Fibonacci sequence. (This is an open-ended problem, and answers will vary according to the particular direction you take.)

Appendix A

The Method of Induction

1 The method

In this appendix, we briefly describe an extremely important proof technique called *proof by induction*. It will be of use in solving some of the problems in the book. We shall first define the term *proposition*.

• A *proposition* is simply a statement or claim; it asserts something, and it could be true or false. Thus, "$10 > 7$" and "$2 > 3$" are propositions (the first one is true, the second is false); "$2^n > n^2$ for all integers $n \geq 5$" is another proposition (true); and so is "the number $2^{2^n} + 1$ is prime for all positive integers n" (false).

Suppose a proposition is to be proved for all positive integers n; say, something like: "For all integers n, the sum $1+2+\cdots+n$ equals $n(n+1)/2$". There are several ways to prove this statement, but here is how it is done via induction.

Let $\mathcal{P}(n)$ denote the proposition "$1+2+\cdots+n = n(n+1)/2$". Then $\mathcal{P}(1)$ is true (it asserts that $1 = 1$, which is true), and so is $\mathcal{P}(2)$, which asserts that $1+2 = \frac{1}{2}(2 \times 3)$ or $3 = 3$. So $\mathcal{P}(n)$ is true for $n = 1$ and $n = 2$.

We shall now assume that $\mathcal{P}(k)$ is true for *some* positive integer k. This means that

$$1+2+\cdots+k = \frac{k(k+1)}{2} \quad \text{for some } k.$$

Adding $k+1$ to both sides of this equality, on the left side we get $1+2+\cdots+k+(k+1)$, and on the right side we get

$$\frac{k(k+1)}{2} + (k+1) = (k+1) \cdot \left(\frac{k}{2} + 1\right) = \frac{(k+1)(k+2)}{2},$$

after simplification. This means that

$$1 + 2 + \cdots + k + (k+1) = \frac{(k+1)(k+2)}{2},$$

which is just the proposition $\mathcal{P}(k+1)$. So if $\mathcal{P}(k)$ is true, then $\mathcal{P}(k+1)$ too is true. But we already know that $\mathcal{P}(1)$ is true; $\therefore$ $\mathcal{P}(2)$ is true; $\therefore$ $\mathcal{P}(3)$ is true; $\therefore$ $\mathcal{P}(4)$ is true; and so on, *ad infinitum*. That is, $\mathcal{P}(n)$ is true for all positive integers n, as had to be proved.

This is the "method of induction". It is of use in situations where, on the basis of observations ("mathematical experiments"), we have arrived at some hypothesis. This hypothesis may be more a piece of guesswork than anything else. For instance, we may examine the statements

$$1 + 2 = 2^2 - 1, \quad 1 + 2 + 2^2 = 2^3 - 1, \quad 1 + 2 + 2^2 + 2^3 = 2^4 - 1,$$

and be led to the educated guess that $1+2+2^2+\cdots+2^{n-1} = 2^n - 1$ for *all* positive integers n.

In general, suppose that we have a proposition $\mathcal{P}(n)$ whose truth we wish to show ($\mathcal{P}(n)$ is a proposition concerning an arbitrary integer n). Also suppose that the following holds good:

(a) $\mathcal{P}(1)$ is true (and readily verified); and

(b) given that $\mathcal{P}(n)$ is true, we may also show $\mathcal{P}(n+1)$ to be true; that is we have the logical implication

$$\text{truth of } \mathcal{P}(n) \text{ implies truth of } \mathcal{P}(n+1).$$

In this scenario, we may confidently assert that $\mathcal{P}(n)$ is true for all positive integers n. This is so because the truth of $\mathcal{P}(1)$ implies that of $\mathcal{P}(2)$, which implies the truth of $\mathcal{P}(3)$, which in turn implies the truth of $\mathcal{P}(4)$, and so on, *ad infinitum*.

In writing out a proof based on the principle of induction, the verification of steps (a) and (b) must be shown clearly. Step (a) is called *anchoring* the induction, while (b) is the *inductive step*.

Sometimes we may need to anchor the induction at an integer greater than 1; for instance, in showing that $2^n \geq n^2$ for all integers $n \geq 5$. In this case, the steps given above are replaced by: (a) verify that $\mathcal{P}(k)$ is true for some positive integer k; and (b) show that for $n \geq k$, the truth of $\mathcal{P}(n)$ implies the truth of $\mathcal{P}(n+1)$.

2 Solved examples

Example 1 We wish to show that for all positive integers n, we have

$$1^2 + 2^2 + 3^2 + \cdots + n^2 = \frac{n(n+1)(2n+1)}{6}.$$

Let $\mathcal{P}(n)$ denote the assertion to be proved. Then $\mathcal{P}(1)$ is true (it asserts that $1^2 = (1 \cdot 2 \cdot 3)/6$ or $1 = 1$); so is $\mathcal{P}(2)$, which asserts that $1^2 + 2^2 = (2 \cdot 3 \cdot 5)/6$ or $5 = 5$. So the induction has been anchored.

Now, assume that $\mathcal{P}(k)$ is true for some positive integer k; then

$$1^2 + 2^2 + 3^2 + \cdots + k^2 = \frac{k(k+1)(2k+1)}{6}.$$

Adding $(k+1)^2$ to both sides, we get $1^2 + 2^2 + 3^2 + \cdots + (k+1)^2$ on the left, and

$$\begin{aligned}\frac{k(k+1)(2k+1)}{6} + (k+1)^2 &= (k+1) \cdot \left(\frac{2k^2+k}{6} + (k+1)\right) \\ &= (k+1) \cdot \frac{2k^2+7k+6}{6} \\ &= \frac{(k+1)(k+2)(2k+3)}{6}\end{aligned}$$

on the right. This shows that the truth of $\mathcal{P}(k)$ implies the truth of $\mathcal{P}(k+1)$. Since the anchoring has been done, this establishes the truth of $\mathcal{P}(n)$ for all integers $n \geq 1$. □

Example 2 We wish to show that the quantity $n(n+1)(n+2)$ is a multiple of 6 for all positive integers n. We proceed as follows.

Let $f(n) = n(n+1)(n+2)$. Then $f(1) = 6$, which is a multiple of 6. So the stated proposition is true for $n = 1$; the induction has been anchored. Next, observe that

$$\begin{aligned}f(n+1) - f(n) &= (n+1)(n+2)(n+3) - n(n+1)(n+2) \\ &= 3(n+1)(n+2).\end{aligned}$$

So the difference between $f(n+1)$ and $f(n)$ is a multiple of 3. Indeed it is a multiple of 6, because one out of $n+1$ and $n+2$ is even, implying that the product $(n+1)(n+2)$ is even. This means that $f(n+1) = f(n)+$ some multiple of 6. Therefore, if $f(n)$ is a multiple of 6, then so is $f(n+1)$. We have now succeeded in carrying out the inductive step. The proof is now complete; we conclude that $n(n+1)(n+2)$ is a multiple of 6 for all positive integers n. $\square$

Example 3 We wish to show, via induction, that $2^n > n^2$ for integers $n \geq 5$.

Let $\mathcal{P}(n)$ denote the statement that $2^n > n^2$. Observe that $\mathcal{P}(n)$ is false for $n = 1$, 2, 3 and 4, but is true for $n = 5$. So we anchor the induction at $n = 5$.

Now, suppose that $2^k > k^2$ for some integer $k \geq 5$; then we have $2^k > k^2$, $\therefore\ 2^{k+1} > 2k^2$. But $2k^2 > (k+1)^2$ for all integers $k \geq 3$; for this inequality is equivalent to $k^2 - 2k > 1$, or $(k-1)^2 > 2$, or $k > \sqrt{2}+1$, that is, $k \geq 3$ since k is an integer. So if $k \geq 5$, then we certainly have $2k^2 > (k+1)^2$, and so we get $2^{k+1} > (k+1)^2$. This proves the inductive step, and the result is established. $\square$

Exercises

A–1 Show that 9 is a divisor of $2^{2n} - 3n - 1$ for all integers $n > 0$.

A–2 Show that that for all integers $n > 0$,

$$\frac{1}{3} + \frac{1}{15} + \frac{1}{35} + \cdots + \frac{1}{(2n)^2 - 1} = \frac{n}{2n+1}.$$

(The denominators are $2^2 - 1 = 3$, $4^2 - 1 = 15$, $6^2 - 1 = 35$, and so on.)

A–3 Show that $2^n > n^3$ for integers $n \geq 10$.

A–4 Show, using induction, that the sum of the interior angles of a convex n-sided polygon is $(n-2)180°$.

A–5 Show that $a^{2n} - b^{2n}$ has $a + b$ as a factor.

Appendix B

Solutions

Chapter 3

3.1.1 The sequence of 1st differences is $\langle 4, 6, 8, 10, 12, \ldots, \rangle$ and the sequence of 2nd differences is $\langle 2, 2, 2, 2, \ldots, \rangle$, a constant sequence.

3.1.2 The difference sequences are, respectively:

(i) $\langle 1, 7, 19, 37, 61, 91, \ldots, \rangle$.

(ii) $\langle 6, 12, 18, 24, 30, \ldots, \rangle$.

(iii) $\langle 6, 6, 6, 6, \ldots, \rangle$.

The sequence of 3rd differences is a constant sequence.

3.1.3 The sequence of 4th differences is $\langle 24, 24, 24, 24, \ldots, \rangle$; it is reached at the 4th stage.

3.1.4 A constant sequence is reached at the 5th stage; it is equal to $\langle 120, 120, 120, 120, \ldots \rangle$.

3.1.5 For the sequence $\{n^k\}$, a constant sequence is reached at the k th stage.

3.1.6 The pattern is the following: the constant for $\{n^k\}$ is the product $1 \times 2 \times 3 \times \cdots k$, which is usually denoted by the symbol $k!$. Thus, the constant for $\{n^6\}$ is $6! = 1 \times 2 \times 3 \times 4 \times 5 \times 6 = 720$.

3.2.1 To obtain the expression for $D\left(\{n^4\}\right)$, we use the identity $(n+1)^4 - n^4 = 4n^3 + 6n^2 + 4n + 1$. Next, we subtract $(4n^3 + 6n^2 + 4n + 1)$ from $\left(4(n+1)^3 + 6(n+1)^2 + 4(n+1) + 1\right)$ and obtain the expression for $D_2\left(\{n^4\}\right)$; and so on.

3.2.2 The answer is

$$\begin{aligned} D_1(A) &= \{5n^4 + 10n^3 + 10n^2 + 5n + 1\}, \\ D_2(A) &= \{20n^3 + 60n^2 + 70n + 30\}, \\ D_3(A) &= \{60n^2 + 180n + 150\}, \\ D_4(A) &= \{120n + 240\}, \\ D_5(A) &= \{120\}. \end{aligned}$$

The computations are done as described above.

3.2.3 Write $F(n)$ as $an^k + \cdots$; then $F(n+1) = a(n+1)^k + \cdots$, so $F(n+1) - F(n) = a(n+1)^k - an^k + \cdots$, where the dots indicate more such terms. Now, when we expand everything out using the Binomial Theorem, the k th powers vanish and the highest degree term left is a $(k-1)$ th power. (Please check this for the case $k = 3$.)

3.2.4 Denote the sequence $A+B$ by C; then $C(n) = A(n) + B(n)$ and the n th term of $D(C)$ is $C(n+1) - C(n)$, which can be written as

$$\{A(n+1) - A(n)\} + \{B(n+1) - B(n)\}.$$

It follows that $D(A+B) = D(A) + D(B)$. The formula for $D(A-B)$ can be worked out in a similar manner.

3.2.5 We need an expression for

$$A(n+1)B(n+1) - A(n)B(n).$$

Adding and subtracting the term $A(n+1)B(n)$, we rewrite the expression as

$$A(n+1)\{B(n+1) - B(n)\} + B(n)\{A(n+1) - A(n)\}.$$

Write P for $D(A)$, Q for $D(B)$, and R for $D(AB)$; then

$$R(n) = A(n+1)Q(n) + B(n+1)P(n),$$

and this is the required formula.

Chapter 4

4.4.1 The generating formulas are listed alongside the sequences.

(i) $\left\langle 1, 3, 7, 13, \ldots \right\rangle$; formula: $P(n) = n^2 + n + 1$.

(ii) $\langle 3, 8, 19, 36, \ldots \rangle$; formula: $Q(n) = 3n^2 + 2n + 3$.

(iii) $\langle 0, 9, 30, 75, \ldots \rangle$; formula: $R(n) = 2n^3 + 7n$.

(iv) $\langle 0, 2, 24, 108, 320, \ldots \rangle$; formula: $S(n) = n^4 + n^3$.

($n = 0$ gives the 1st term listed, $n = 1$ gives the 2nd term, ...)

4.4.2 We need to find two formulas that fit the sequence P:

$$P = \langle 1, 3, 9, \ldots \rangle.$$

As we have been provided with just three terms of the given sequence, we are free to extend the sequence in any manner convenient to us! The array of differences for P is displayed below.

$$\begin{array}{ccccc} 1 & & 3 & & 9 \\ & 2 & & 6 & \\ & & 4 & & \end{array}$$

We now choose to continue the array in two different ways as shown below. Note the pattern in each case; a constant sequence is reached in the 3rd row in the 1st case, and in the 4th row in the 2nd case. We name the sequences that produce these arrays Q and R, respectively.

<u>Array of differences for Q</u>

$$\begin{array}{ccccccc} 1 & & 3 & & 9 & & 19 \\ & 2 & & 6 & & 10 & \\ & & 4 & & 4 & & \end{array}$$

<u>Array of differences for R</u>

$$\begin{array}{ccccccccc} 1 & & 3 & & 9 & & 25 & & 57 \\ & 2 & & 6 & & 16 & & 32 & \\ & & 4 & & 10 & & 16 & & \\ & & & 6 & & 6 & & & \end{array}$$

You may wonder how we build up these arrays. This is easy—*we simply start from the bottom!* That is, we decide in advance that we want, say, the 3rd row (in the case of Q) to be a row of constants. Since we already have one entry in that row ($= 4$), we fill the row with 4s and proceed to build the array upwards, going in the reverse direction as it were. (Please work out the details on your own.)

The two sequences Q and R we thus obtain are:

$$Q = \left\langle 1,\ 3,\ 9,\ 19,\ 33,\ 51,\ \dots \right\rangle,$$
$$R = \left\langle 1,\ 3,\ 9,\ 25,\ 57,\ \dots, \right\rangle.$$

We now apply the usual technique and obtain the following formulas:

$$Q(n) = 2n^2 + 1,$$
$$R(n) = n^3 - n^2 + 2n + 1.$$

To verify that their correctness, we substitute a few sample values for n: $Q(0) = 1$, $Q(1) = 3$, $Q(2) = 8 + 1 = 9$, $Q(3) = 18 + 1 = 19$, $\dots$; $R(0) = 1$, $R(1) = 1 - 1 + 2 + 1 = 3$, $R(2) = 8 - 4 + 4 + 1 = 9$, $R(3) = 27 - 9 + 6 + 1 = 25$, $\dots$. Everything seems alright, and we have two different formulas as required. (As you will anticipate, more such formulas can be worked out merely by building up the array differently.)

4.9.1 The required sum is

$$\left(\frac{n(n+1)}{2}\right)^2 \quad \text{[which also equals } (1+2+3+\cdots+n)^2\text{]}.$$

To show that the answer is correct, it is sufficient if we verify that

$$\left(\frac{n+1)(n+2)}{2}\right)^2 - \left(\frac{n(n+1)}{2}\right)^2 = (n+1)^3.$$

The left side is a difference of two squares, so we use the formula $a^2 - b^2 = (a-b)(a+b)$. The difference and sum of the two factors are

$$\left(\frac{(n+1)(n+2)}{2} - \frac{n(n+1)}{2}\right) = (n+1),$$
$$\left(\frac{(n+1)(n+2)}{2} + \frac{n(n+1)}{2}\right) = (n+1)^2,$$

so the product does simplify to $(n+1)^3$, as required.

4.9.2 The required sum is

$$\frac{n(n+1)(2n+1)(3n^2+3n-1)}{30}.$$

(Check: For $n = 3$, the sum equals $1 + 16 + 81 = 98$. The formula gives the sum as

$$\frac{3 \times 4 \times 7 \times 35}{30} = 98,$$

so the formula is verified in this instance.)

4.9.3 The required sum can be written as

$$1^2 + 3^2 + 5^2 + \cdots + (2n-1)^2$$

and this equals $A - B$, where

$$A = 1^2 + 2^2 + 3^2 + 4^2 + \cdots + (2n-1)^2 + (2n)^2,$$
$$B = 2^2 + 4^2 + 6^2 + 8^2 + \cdots + (2n-2)^2 + (2n)^2.$$

(Note that B is the sum of the even squares till $(2n)^2$.) Now, we already know the formula for the sum $1^2 + 2^2 + 3^2 + \cdots + n^2$; it is

$$\frac{n(n+1)(2n+1)}{6}.$$

It follows that

$$A = \frac{(2n)(2n+1)(4n+1)}{6}$$
$$B = 4\left(\frac{n(n+1)(2n+1)}{6}\right).$$

Thus, the required sum is

$$A - B = \frac{n(2n-1)(2n+1)}{3}.$$

(Try the value $n = 4$. The actual sum is $1 + 9 + 25 + 49 = 84$, while the formula gives the answer as

$$\frac{4 \times 7 \times 9}{3} = 84.)$$

4.9.4 We need a formula for the sum

$$1^2 + 4^2 + 7^2 + \cdots + (3n-2)^2.$$

We shall follow the usual procedure. The constant for the given sequence (call it A) turns out to be 18, and it is reached after three repetitions of the process, so we subtract $\{3n^3\}$ from A; the result is the sequence B, where

$$B = \langle -2,\ -7,\ -15,\ -26,\ -40,\ \dots, \rangle.$$

Repeating the procedure for B, we find that the constant for this sequence is -3, and it is reached at the 2nd stage. Since -3 is equal to $-1/2 \times 6$, we subtract $\{-3n^2/2\}$ from B. The result is the sequence C, where

$$C = \left\langle -\frac{1}{2},\ -1,\ -\frac{3}{2},\ -2,\ \dots, \right\rangle,$$

which is clearly $\{-n/2\}$. It follows that

$$A = \left\{ 3n^3 - \frac{3}{2}n^2 - \frac{n}{2} \right\},$$

which simplifies to $A = \{n(6n^2 - 3n - 1)/2\}$. Therefore,

$$1^2 + 4^2 + 7^2 + \cdots + (3n-2)^2 = \frac{n(6n^2 - 3n - 1)}{2}.$$

Chapter 5

5.2.1 The array of differences for $\{4^n\}$ consists entirely of numbers of the form $3^a 4^b$, where a and b are non-negative integers; no number occurs twice in the array. The starting numbers of the rows are the powers of 3.

Similarly, the array of differences for $\{5^n\}$ consists entirely of numbers of the form $4^a 5^b$, where a and b are non-negative integers; as earlier, no number occurs twice in the array. The starting numbers of the rows are 1, 4, 16, 64, ..., that is, the powers of 4.

5.2.2 The results are displayed below.

(i) $\{3^n + n\}$: the component $\{n\}$ gets erased while taking the 2nd differences. Thus, in the sequence of 2nd differences ($= \langle 4, 12, 36, \dots, \rangle$), each term is three times the one preceding it.

(ii) $\{3^n + n^2\}$: the $\{n^2\}$ component gets erased while taking the 3rd differences. Thus, in the sequence of 3rd differences ($= \langle 8, 24, 72, \dots \rangle$), each term is three times the one preceding it.

(iii) $\{3^n + n^2 + n\}$: the same as in (ii).

(iv) $\{2^n + n^3\}$: the $\{n^3\}$ component gets erased while taking the sequence of 4th differences.

(v) $\{3^n + 2^n\}$: both components are exponential, so neither one gets erased! The successive arrays look increasingly complicated in this case, although they do have a simple (but hidden) structure.

5.3.1 The sequence of 1st differences is

$$D\{a^n\} = \langle a-1, a(a-1), a^2(a-1), \ldots \rangle,$$

so $D_1(\{a^n\}) = \{a^n(a-1)\}$. We find similarly that $D_2(\{a^n\}) = \{a^n(a-1)^2\}$, $D_3(\{a^n\}) = \{a^n(a-1)^3\}$, and so on.

5.3.2 We have, $(-1)^{n+1} - (-1)^n = (-2) \cdot (-1)^n$. Therefore,

$$\begin{aligned} D_1(\{(-1)^n + n^3\}) &= \{(-2) \cdot (-1)^n + 3n^2 + 3n + 1\}, \\ D_2(\{(-1)^n + n^3\}) &= \{4 \cdot (-1)^n + 6n + 6\}, \\ D_3(\{(-1)^n + n^3\}) &= \{(-8) \cdot (-1)^n + 6\}, \\ D_4(\{(-1)^n + n^3\}) &= \{16 \cdot (-1)^n\}. \end{aligned}$$

5.3.3 Following the methods used in 5.3.1 and 5.3.2, we see that the successive difference arrays of $\{4^n\}$ are $\{3 \cdot 4^n\}$, $\{3^2 \cdot 4^n\}$, $\{3^3 \cdot 4^n\}$, ... It follows that the numbers occurring in the array of differences are all divisors of the powers of 12. (They are also, naturally, also divisors of the powers of 6, but not every divisor of a power of 6 appears in the array.)

5.3.4 Following the methods used above, we see that the successive difference arrays of $\{5^n\}$ are $\{4 \cdot 5^n\}$, $\{4^2 \cdot 5^n\}$, $\{4^3 \cdot 5^n\}$, ... It follows that the numbers occurring in the array of differences are all divisors of the powers of 20.

Chapter 6

6.1.1 The number of lines is $L(n) = n(n-1)/2$. (To see why this is so, note that each point has $n-1$ lines passing through it, one corresponding to each of the other $n-1$ points. Each of these lines is counted precisely *twice*. How do you complete the argument from here?)

6.1.2 The number of points of intersection is $P(n) = n(n-1)/2$. (Once the formula has been worked out, it is easy to see why it must be correct. Here is one way of seeing why: there are n lines in all and each line intersects each of the others, so there are $n-1$ points of intersection on each line. Therefore, we multiply n by $n-1$, giving $n(n-1)$ points in all. However, we have counted each point *twice* in this process, so we divide the result by 2. The result is the formula $n(n-1)/2$.)

6.1.3 The number of regions is

$$R(n) = \frac{n^2 + n + 2}{2}.$$

The formula can be proved in an elegant manner. Consider the situation when $(n-1)$ lines have been drawn, producing $R(n-1)$ regions, and examine what happens when the nth line is drawn: it meets each of the other $(n-1)$ lines and is therefore, divided into n segments. Each segment marks the edge of a newly-created region, so n new regions get created. Therefore, $R(n) = R(n-1) + n$, or $R(n) - R(n-1) = n$. We now obtain

$$\begin{aligned} R(n) &= (R(n) - R(n-1)) + (R(n-1) - R(n-2)) \\ &\quad + \cdots + (R(3) - R(2)) + (R(2) - R(1)) + R(1) \\ &= n + (n-1) + \cdots + 3 + 2 + 2 \\ &= (n + (n-1) + \cdots + 3 + 2 + 1) + 1 \\ &= \frac{n(n+1)}{2} + 1 \;=\; \frac{n^2 + n + 2}{2}. \end{aligned}$$

Therefore, $R(20) = (20 \times 21)/2 + 1 = 211$. This answers the question asked in Chapter 1.

6.1.4 The number of pieces is equal to

$$P(n) = \frac{n^4 - 6n^3 + 23n^2 - 18n + 24}{24}.$$

For example, for $n = 4$, the formula gives the number of pieces as $(256 - 384 + 368 - 72 + 24)/24 = 8$, which is correct. Note the complexity of the answer!

For those who are familiar with counting arguments and the use of binomial coefficients, here is a rather sophisticated way of proving the result. The argument involves "disentangling" the chords from one another and listing how the regions are created as each additional chord is drawn. At the start, when no chords have been drawn, there is just one region (the whole circle). At any intermediate stage, let a new chord be drawn; let it be cut into k pieces by the chords already drawn. Then the number of new regions created is $k+1$. This can be written as 1+ # intersections the chord has with the chords drawn earlier. Arguing thus, we deduce that in the end the number of regions is equal to 1+ # chords + # points of intersection, which in turn can be written as

$$1+\binom{n}{2}+\binom{n}{4}.$$

This simplifies to the formula displayed above.

6.1.5 The number of points of intersection is

$$\begin{aligned} I(n) &= n+3\cdot\frac{n(n-1)(n-2)(n-3)}{24} \\ &= \frac{n(n^3-6n^2+11n+2)}{8}. \end{aligned}$$

As in the previous case, there is a beautiful way of proving the formula, but the proof is rather difficult and requires some familiarity with the use of binomial coefficients.

Here are the details. For every four points chosen from the set, there correspond precisely *three* points of intersection. (If the points are A, B, C and D, then the points of intersection are $AB\cap CD$, $AC\cap BD$ and $AD\cap BC$.) The number of sets of four points from a collection of n points is by definition $\binom{n}{4}$, so the total number of points of intersection is $3\cdot\binom{n}{4}$. To this, we add n to account for the original n points. Simplifying, we obtain the formula given above.

Chapter 7

7.7.1 The forms $5k$, $5k\pm1$, $5k\pm2$ exhaust all the integers. Squaring these expressions, we get $25k^2$, $25k^2\pm10k+1$, $25k^2\pm20k+4$, respectively, and these are of the forms $5k'$, $5k'+1$, $5k'+4$.

7.7.2 The forms $7k$, $7k\pm1$, $7k\pm2$ and $7k\pm3$ exhaust all the integers, and their squares are of the forms $7k'$, $7k'+1$, $7k'+4$, $7k'+2$, respectively. Observe that the form $7k'-1$ (which is the same as $7k'+6$) does not occur in this list.

7.7.3 The forms $4k \pm 1$ exhaust all the odd numbers, and their squares are of the form $8k' + 1$. This means that odd squares leave a remainder of 1 when divided by 8.

7.7.4 The forms $8k \pm 1$, $8k \pm 3$ exhaust the odd numbers, and their squares have the forms $16k' + 1$, $16k' + 9$, respectively. So the possible remainders left when odd squares are divided by 16 are 1 and 9.

7.7.5 $73 = 8^2 + 3^2$, $109 = 10^2 + 3^2$, $449 = 20^2 + 7^2$, $1009 = 28^2 + 15^2$.

7.7.6 Suppose that n can be so written, say $n = a^2 - b^2 = (a - b)(a + b)$. Observe that the numbers $a \pm b$ have the same parity (they are both odd or both even), so either n is odd, or if it is even then it is a multiple of 4. Thus, a necessary condition for n to be of this form is that if it is even, then it should be a multiple of 4. In fact, these conditions are sufficient as well: if n has this property, then write n as a product cd where c and d are both odd or both even and $c > d$. Now, let $a = (c + d)/2$ and $b = (c - d)/2$; then $n = a^2 - b^2$. (Example: Let $n = 80 = 20 \times 4$; this gives $a = 24/2 = 12$ and $b = 16/2 = 8$; so $80 = 12^2 - 8^2$.)

Chapter 8

8.5.1 The forms $3k$, $3k \pm 1$ exhaust all the integers, and their cubes are of the forms $9k'$, $9k' \pm 1$.

8.5.2 The cubes of the numbers $0, \pm 1, \pm 2, \pm 3, \pm 4, \pm 5, \pm 6$ are $0, \pm 1, \pm 8, \pm 27, \pm 64, \pm 125, \pm 216$, respectively, and these numbers leave the following remainders on division by 13: $0, \pm 1, \pm 5, \pm 1, \pm 1, \pm 5, \pm 5$, respectively. It follows that the remainders left when cubes indivisible by 13 are divided by 13 are ± 1 or ± 5.

8.5.3 The same method followed in Problem 8.5.2 may be followed here also. The details are left to the reader.

8.5.4 This is a matter of routine verification, and we leave the task to the reader.

8.5.5 Clearly N is divisible by the primes 2, 3 and 5. Let $N = 2^a 3^b 5^c$. Then $N/2 = 2^{a-1} 3^b 5^c$, so for $N/2$ to be a square, $a - 1, b, c$ must be even. Similarly, $a, b - 1, c$ must be multiples of 3, and $a, b, c - 1$ must be multiples of 5. Therefore, a is an odd multiple of 15; b is a multiple of 10 and leaves a remainder of 1 on division by 3; and c is a multiple of 6 and leaves a remainder of 1 on division by 5. The obvious choices are $a = 15$, $b = 10$, $c = 6$. So the smallest possible positive number with the property is the number $2^{15}3^{10}5^{6}$, which equals 30233088000000.

Chapter 10

10.8.1 Let $S = 1 + 3 + 3^2 + \cdots + 3^n$; then $3S = 3 + 3^2 + 3^3 + \cdots + 3^n + 3^{n-1}$, so by subtraction, $2S = 3^{n+1} - 1$, and $S = (3^{n+1} - 1)/2$.

10.8.2 If S represents the given sum, then $4S = 4 + 4^2 + 4^3 + \cdots + 4^{n+1}$, so by subtraction, $3S = 4^{n+1} - 1$, and this yields the required result.

10.8.3 Following the method used in Problems 10.8.1 and 10.8.2, we find that the required sum is $(a^{n+1} - 1)/(a - 1)$.

10.8.4 The remainders left when 1, 2, 4, 8, 16, 32, 64, ... are divided by 3 are 1, 2, 1, 2, 1, 2, 1, ..., and we see that the sequence 1, 2 repeats indefinitely. This is a genuine pattern, as may be checked by noting that if $2^n = 3k + 1$, then $2^{n+1} = 6k + 2$, which leaves a remainder of 2 on division by 3; whereas if $2^n = 3k + 2$, then $2^{n+1} = 6k + 4$, which leaves a remainder of 1 on division by 3. So a 1 is followed by a 2, and a 2 in turn is followed by a 1.

10.8.5 The remainders left when 1, 2, 4, 8, 16, 32, 64, ... are divided by 5 are: 1, 2, 4, 3, 1, 2, 4, 3, ...; we see that the cycle 1, 2, 4, 3 repeats indefinitely. This is genuine; for if $2^n = 5k+1$ then $2^{n+1} = 10k + 2 = 5k_1 + 2$; next, $2^{n+2} = 10k_1 + 4 = 5k + 2 + 4$; and $2^{n+3} = 10k_2 + 8 = 5k_3 + 3$; and finally, $2^{n+4} = 10k_3 + 6 = 5k + 4 + 1$. So we come back to a remainder of 1 after every 4 steps.

10.8.6 The sequence of remainders obtained when the powers of 2 are divided by 7 is 1, 2, 4, 1, 2, 4, ...; the cycle 1, 2, 4 repeats indefinitely. It may be verified as in Problems 10.8.4 and 10.8.5 that the pattern is genuine.

Observe that a remainder of 6 does not appear in the cycle. This implies that $2^n + 1$ is not divisible by 7 for any integer $n > 0$.

10.8.7 We consider the equations $2^a - 5^b = \pm 1$, where a, b are positive integers.

Suppose that $2^a - 5^b = 1$ for some integers $a, b > 0$. This means that 2^a is of the form $5k+1$, so from Problem 10.8.5 it follows that a is a multiple of 4. Therefore, a is even, and so (Problem 10.8.4) 2^a is of the form $3k'+1$. Thus, we get $5^b + 1 = 3k' + 1$, or $5^b = 3k'$, implying that a power of 5 is divisible by 3; which is absurd, as a power of 5 can have only 5 as a prime factor. So the equation cannot hold good.

The equation $2^a - 5^b = -1$ has the obvious solution $a = 2$, $b = 1$. Suppose that there exists a solution with $a \geq 3$. Then 2^a is a multiple of 8, so 5^b is of the form $8k+1$. Now the remainders left when the powers of 5 are divided by 8 are; 5, 1, 5, 1, . . . : the cycle $(5, 1)$ repeats indefinitely. The remainder of 1 occurs when the exponent is even, so b is even. Next, observe that 2^a is of the form $5k-1$, that is, of the form $5k+4$, and using the cycle $(2, 4, 3, 1)$ we deduce that a is even. So both a and b are even, which means that 2^a and 5^b are both squares. But there is no pair of non-zero squares with a difference of 1! So this case too is impossible. It follows that the only solution of $2^a - 5^b = \pm 1$ is $a = 2$, $b = 1$.

10.8.8 Let us consider the equation $2^a - 7^b = \pm 1$.

The equation $2^a - 7^b = 1$ has the obvious solution $a = 3$, $b = 1$. Suppose that there exists a solution with $a \geq 4$, $b \geq 2$. Then 2^a is a multiple of 16, so 7^b is of the form $16k-1$. Now the remainders left when the powers of 7 are divided by 16 are 7, 1, 7, 1, 7, 1, . . . ; the cycle $(7, 1)$ repeats. Observe that a remainder of 15 never occurs. It follows that $7^b + 1$ can never be of the form $16k$, so this case is impossible.

If $2^a - 7^b = -1$ for some integers $a, b > 0$, then $2^a + 1 = 7^b$, so $2^a + 1$ is divisible by 7. However, we know that this is cannot be (Problem 10.8.6). So this case too is impossible. We conclude that in the mixed powers-of-2-and-7 sequence, the only pair of consecutive integers is $(7, 8)$.

10.8.9 Suppose that $1000 = a+(a+1)+(a+2)+\cdots+(a+b)$, where $a, b > 0$. Then $1000 = (b+1)(2a+b)/2$, so $2000 = (b+1)(2a+b)$. Now 2000 may be written as a product of two positive integers of opposite parity in two ways: 16×125 and 25×80. The 1st way gives $b = 15$, $a = 55$ and the 2nd way gives $b = 24$, $a = 28$. So $2000 = 55+56+57+\cdots+69+70$, and also $2000 = 28+29+30+\cdots+51+52$.

Similarly, 3000 may be written as 3×1000, 5×600 8×375, 15×200, 24×125, 25×120 and 40×75. These possibilities give

(a, b) equal to $(499, 20)$, $(298, 4)$, $(184, 7)$, $(93, 14)$, $(51, 23)$, $(48, 24)$ and $(18, 39)$, respectively. Each of these yields a solution; e.g., $(a, b) = (51, 23)$ corresponds to $1500 = 51 + 52 + 53 + \cdots + 73 + 74$.

10.8.10 This may be equivalently be phrased as: *Which positive integers can be written as differences of two squares?* The answer is: all odd numbers, and all multiples of 4. (See Problem 7.7.6.)

Chapter 11

11.4.1 We can argue by contradiction. Suppose that the given AP contains a last prime, say $3N - 1$. Consider the number K given by

$$K = 3(2 \times 5 \times 8 \times \cdots \times (3N - 1)) - 1.$$

Then K belongs to the same AP, and by construction is not divisible by any of the numbers 2, 5, 8, ..., $3N - 1$. Since K has prime factors, these primes do not belong to the list 2, 5, 8, ..., $3N - 1$. Now the primes dividing K cannot all be of the form $3s + 1$, because the product of any number of such primes is itself of this form, contradicting the fact that K itself is of the form $3s - 1$. So K has some prime factor which belongs to the given AP, but it cannot be a number occurring prior to $3N - 1$. Therefore, the AP must have a prime after $3N - 1$.

11.4.2 Here the AP consists of numbers of the form $7s - 1$. We repeat the argument of Problem 11.4.1 verbatim: assuming that the largest prime in the given AP is $7N - 1$, we consider the number K given by

$$K = 7(6 \times 13 \times 20 \times \cdots \times (7N - 1)) - 1.$$

As earlier, we argue that this must possess a prime factor which belongs to the given AP but is larger than $7N - 1$. This proves that the AP contains infinitely many primes.

11.4.3 Each odd number may be written as $6k$ plus 1, 3 or 5. So the odd numbers, relative to the divisor 6, leave the following remainders: 1, 3, 5, 1, 3, 5, 1, 3, 5, ... Since the cycle length is 3, a '3' occurs in any stretch of 3 consecutive members of this sequence. This justifies the claim.

A similar proof holds good for the case of three consecutive numbers.

11.4.4 If $a, a+d, a+2d$ are all prime, with $a > 3$, then a and $a+d$ are odd and therefore d is even. Suppose that a is of the form $3k+1$. If d is of the form $3k'+1$, then $a+2d = 3(k+2k'+1)$ is a multiple of 3 and is therefore non-prime because $k+2k'+1 > 1$; and if d is of the form $3k'+2$, then $a+d = 3(k+k'+1)$ is a multiple of 3 and is therefore non-prime because $k+k'+1 > 1$. So d must be a multiple of 3. Since d is even, this means that d is a multiple of 6.

11.4.5 There are many such triples; e.g., 3, 43, 73; 1, 31, 61; 11, 41, 71; and so on.

11.4.6 This follows from a simple result on divisibility: *If a divides b, then R_a divides R_b.* Thus, R_2 divides R_4 $(1111/11 = 101)$ and R_3 divides R_6 $(111111/111 = 1001)$. The reason behind this claim should be obvious. It follows that if n is composite, then R_n is composite (if p is a prime divisor of n then R_p is a divisor of R_n).

11.4.7 We first observe that

$$R_{p-1} = \underbrace{111\ldots1}_{(p-1)\text{ 1's}} = \frac{10^{p-1}-1}{9}.$$

Suppose that $p > 5$ is prime. Then 10 is not a multiple of p, so Fermat's little theorem applies: p is a divisor of $10^{p-1}-1$. Since $R_{p-1} = (10^{p-1}-1)/9$ and p and 9 are coprime, the presence of p in the factorization of $10^{p-1}-1$ is not "lost" after division by 9; so p is a divisor of R_{p-1} as well.

The argument will not work for $p = 2$ and 5, because 2 and 5 are divisors of 10, and it will not work for $p = 3$ because 3 is a divisor of 9; hence, the restriction on p.

11.4.8 Since $496 = 16 \times 31$, the proper divisors of 496 are 1, 2, 4, 8, 16, 31, 62, 124, 248. The sum of the divisors is

$$16 + (1+2+4+8) \times (1+31) = 16 + (15 \times 32) = 16 \times 31.$$

Similarly, $8128 = 64 \times 127$, so the proper divisors of 8128 are 1, 2, 4, 8, 16, 32, 64, 127, 254, 508, 1016, 2032, 4064. Their sum is

$$64 + (1+2+4+8+16+32) \times (1+127) = 64 + (63 \times 128) = 64 \times 127.$$

11.4.9 The proof below is given for the Fermat numbers F_4 and F_7, but the same idea works for any F_m and F_n $(m \neq n)$. Note that $F_7 = 2^{128} + 1$ and $F_4 = 2^{16} + 1$. Since $2^{128} = \left(2^{16}\right)^8$, we see that

$$F_7 = (F_4 - 1)^8 + 1.$$

On simplifying the expression on the right side, we find that $F_7 = k \cdot F_4 + 2$ for some integer k. So if d is a common factor of F_4 and F_7, then d is a divisor of 2; that is, $d = 1$ or 2. Since the Fermat numbers are all odd, d cannot be 2. This means that $d = 1$; that is, 1 is the only common factor of F_4 and F_7. Thus, F_4, F_7 are coprime.

Each Fermat number has its own set of prime factors, and since the Fermat numbers are coprime, the various sets of primes are disjoint. As there are infinitely many Fermat numbers, there must be infinitely many primes.

11.4.10 By definition, the nth Euclidean number E_n is $p_1 p_2 \cdots p_n +$ 1. Here $p_1 = 2$ and p_2, p_3, ..., are all odd. Let the product $p_2 p_3 \ldots p_n$ be written as $2k+1$; then $E_n = 2 \times (2k+1) + 1 = 4k+3$. However, all squares are of the form $4k$ or $4k+1$; so a Euclidean number is never a square.

Chapter 12

12.2.1 Let $T_a : T_b = 1 : 3$; then $3a(a+1) = b(b+1)$ or, multiplying by 4, $3(4a^2 + 4a) = (4b^2 + 4b)$. Let $x = 2a + 1$, $y = 2b + 1$; then $3(x^2 - 1) = y^2 - 1$, so $y^2 - 3x^2 = -2$. The solutions of this equation are given below:

$$(y, x) = (1, 1),\ (5, 3),\ (19, 11),\ (71, 41),\ (265, 153),\ \ldots.$$

The 1st pair yields $a = b = 0$, which we must discard. The other pairs yield

$$(b, a) = (2, 1),\ (9, 5),\ (35, 20),\ (132, 76),\ \ldots$$

Thus, $T_2 = 3T_1$, $T_9 = 3T_5$, $T_{35} = 3T_{20}$, $T_{132} = 3T_{76}$, and so on. The pattern is easy to spot: if (a, b), (a', b'), (a'', b'') are three successive solution pairs, then $a'' = 4a' - a + 1$, $b'' = 4b' - b + 1$.

12.2.2 The Fermat numbers are of the form $2^{2^n} + 1$. For $n > 0$, 2^n is even, so $2^{2^n} + 1$ is of the form $3k + 1$. (See Problem 10.8.4) This means that after the 1st Fermat number, the others are all of the form $3k + 2$.

On the other hand, the triangular numbers are all of the form $3k$ or $3k+1$; for if n is of any of the forms $6k$, $6k+2$, $6k+3$, $6k+5$, then $T_n = n(n+1)/2$ is a multiple of 3, that is of the form $3k'$. And if n is of one of the forms $6k+1$, $6k+4$, then $T_n = (6k+1)(3k+1)$, $(3k+2)(6k+5)$, respectively. When the expressions are opened out, we find that they are of the form $3k'+1$. So T_n is always of one of the forms $3k'$ or $3k'+1$. It follows that the only triangular number which is also a Fermat number is 3.

12.2.3 Observe that $1/T_n = 2/n(n+1) = 2(1/n - 1/(n+1))$. Therefore, the sum $1/T_1 + 1/T_2 + \cdots + 1/T_n$ is equal to

$$2 \cdot \left\{ \left(\frac{1}{1} - \frac{1}{2}\right) + \left(\frac{1}{2} - \frac{1}{3}\right) + \cdots + \left(\frac{1}{n} - \frac{1}{n+1}\right) \right\}.$$

Practically everything cancels out within the big braces; only the 1st and last terms are left (the sum has "telescoped"), so the sum is $2(1/1 - 1/(n+1))$, which simplifies to $2n/(n+1)$.

12.2.4 The nth pentagonal number is $P_n = n(3n-1)/2$. Suppose that P_a is a square, say $P_a = b^2$. Then $a(3a-1) = 2b^2$, so $9a^2 - 3a = 6b^2$; completing the square $(3a - 1/2)^2 = 6b^2 + 1/4$, or $(6a-1)^2 - 24b^2 = 1$. Write $A = 6a - 1$, $B = 2b$; then $A^2 - 6B^2 = 1$, a familiar type of equation. Its solutions are given by $(A, B) = (5, 2)$, $(49, 20)$, $(485, 198)$, $(4801, 1960)$, $\ldots$, and the pattern is easy to see: if (u, v), (u', v'), (u'', v'') are successive solution pairs, then $u'' = 10u' - u$, $v'' = 10v' - v$. The A-values are alternately of the form $6k - 1$ and $6k + 1$. Our interest is in the values of the form $6k - 1$, so the solution pairs of relevance to us are $(A, B) = (5, 2)$, $(485, 198)$, $(47525, 19402)$, $(4656965, 1901198)$, $\ldots$. Since $a = (A+1)/6$ and $b = B/2$, these yield

$$(a, b) = (1, 1),\ (81, 99),\ (7921, 9701), (776161, 950599),\ \ldots.$$

So the pentagonal numbers which are also squares are $P_1 = 1^2$, $P_{81} = 99^2$, $P_{7921} = 9701^2$, $\ldots$. If (a, b), (a', b') and (a'', b'') are successive members in this sequence, then $a'' = 98a' - a - 16$, $b'' = 98b'' - b'$.

12.2.5 Suppose that $P_a = 2P_b$; then $a(3a-1) = 2b(3b-1)$, or after completing the square in the usual way, $(6a-1)^2 = 2(6b-1)^2 - 1$. Write A for $6a - 1$ and B for $6b - 1$; then $A^2 - 2B^2 = -1$, and A, B

are both of the form $6k-1$. The solution pairs of the equation $A^2-2B^2=-1$ are $(A,B)=(1,1),(7,5),(41,29),(239,169),\ldots$. The pairs in which both A and B are of the form $6k-1$ occur at the 3rd, 7th, 11th, ..., places in this sequence (at intervals of 4): $(A,B)=(41,29),(47321,33461),\ldots$. Since $a=(A+1)/6$ and $b=(B+1)/6$, we get

$$(a,b)=(7,5),\ (7887,5577),\ \ldots$$

So $P_7=2P_5$, $P_{7887}=2P_{5577}$, and so on. The numbers grow unmanageably large rather rapidly.

Chapter 13

13.2.1 We have, $1/3+1/4<1$, $1/5+1/6+1/7+1/8<1$, ..., and $1/(2^{n-1}+1)+1/(2^{n-2}+1)+\cdots+1/2^n<1$. Adding all these inequalities, we get $1+1/2+1/3+\cdots+1/2^n<n+1$.

13.2.2 Let s_n denote the sum $1+1/2+1/2^2+\cdots+1/2^n$. Then $2s_n=2+1+1/2+\cdots+1/2^{n-1}$, and by subtraction we get $s_n=2-1/2^n$. Thus s_n is less than 2, but the gap is small, only $1/2^n$, which decreases rapidly as n increases. As n grows without limit, s_n tends to 2.

13.2.3 Proceeding in exactly the same way, we write t_n for the sum $1+1/3+1/3^2+\cdots+1/3^n$. Then $3t_n=3+1+1/3+\cdots+1/3^{n-1}$, so $2t_n=3-1/3^n$. As n grows without limit, $2t_n$ tends to 3; that is, t_n tends to 3/2.

13.2.4 Proceeding as in Problems 13.2.2 and 13.2.3, let a_n denote the sum $1+0.9+0.9^2+\cdots+0.9^n$. Then $0.9a_n=0.9+0.9^2+\cdots+0.9^{n+1}$, so by subtraction we get $0.1a_n=1-0.9^{n+1}$. Thus, as n grows without limit, $0.1a_n$ tends to 1; that is, a_n tends to 10.

13.2.5 Recall that s_k is the sum of the reciprocals of the k-digit numbers which contain no 0s, and that A_k denotes the set of such k-digit numbers (so A_k contains 9^k numbers in all). Corresponding to each number n in A_k, the following 9 numbers occur in A_{k+1}: $10n+1$, $10n+2$, $10n+3$, ..., $10n+9$. Clearly, the sum of their reciprocals is *less* than $1/10n+1/10n+\cdots+1/10n=9/10n=0.9/n$. This shows directly that $s_{k+1}/s_k<0.9$.

13.2.6 Let A_n denote the set of n-digit numbers with no 1s (with all numbers written in base-10). So A_1 consists of the 8 numbers

2, 3, ..., 9, A_2 consists of the $8 \times 9 = 72$ numbers 22, 23, ..., 29, 30, 32, 33, ..., 99. Observe that A_n has $8 \times 9^{n-1}$ numbers, all between 10^{n-1} and 10^n. So the sum of the reciprocals of the numbers in A_n lies between $8 \times 9^{n-1}/10^n$ and $8 \times 9^{n-1}/10^{n-1}$; that is, between $0.8 \times 0.9^{n-1}$ and $8 \times 9^{n-1}$. From Problem 13.2.4, we know that the infinite sum $1 + 0.9 + 0.9^2 + \cdots$ equals 10, so it follows that the sum of the reciprocals of the base-10 numbers which have no 1s lies between 0.8×10 and 8×10; that is, between 8 and 80. This is a wide range, but it is all we need to show that the sum in question is *finite*.

Estimating the sum more accurately is a much trickier matter, and we leave the problem for the student to investigate.

13.2.7 The problem is to estimate the infinite sum $1/1 + 1/11 + 1/111 + \cdots$. Let s_n denote the sum of the 1st n terms of this series. Using the computer, we find that $s_4 = 1.10081818\ldots$, $s_5 = 1.1009081908\ldots$, ..., $s_{10} = 1.100918190736\ldots$. Let a_n denote the infinite sum

$$\frac{1}{\underbrace{111\ldots11}_{(n+1)\ 1s}} + \frac{1}{\underbrace{111\ldots111}_{(n+2)\ 1s}} + \cdots$$

(so the required sum is $s_n + a_n$). Then a_n is less than $1/10^n + 1/10^{n+1} + \cdots$, but more than $9 \times (1/10^{n+1} + 1/10^{n+2} + \cdots)$. That is, a_n lies between $0.00\ldots00999\ldots$ (with n zeroes before the 1st 9) and $0.00\ldots0111\ldots$ (with $(n-1)$ zeroes before the 1st 1). Since $1.10081818 + 0.0000999\ldots$ is $1.10091818\ldots$, and $1.10081818 + 0.000111$ is $1.10092929\ldots$, the required sum certainly lies between 1.100918 and 1.100930. Applying the same reasoning to s_{10}, we find that the required sum lies between $1.100918190836\ldots$, and $1.100918190847\ldots$. By using this method, we can estimate the sum to any desired accuracy.

13.2.8 The task is to estimate the infinite sum $1/1 + 1/2 + 1/11 + 1/12 + 1/21 + 1/22 + 1/111 + \cdots$. Consider the k-digit numbers composed only of 1s and 2s; there are 2^k numbers of this kind, the least being $11\ldots11$ (with k 1s) and the largest $22\ldots22$ (with k 2s). Let s_k be the sum of the reciprocals of these numbers. Here are some initial values of s_k: $s_1 = 1.5$, $S_2 \approx 0.267316$, $s_3 \approx 0.052885$, $s_4 \approx 0.010565$, $s_5 \approx 0.002113$, $s_6 \approx 0.000423$. We find, after some calculation, that $s_2/s_1 \approx 0.178$, $s_3/s_2 \approx 0.1978$, $s_4/s_3 \approx 0.19978$, $s_5/s_4 \approx 0.199975$, ... Observe that for large k the quantity s_{k+1}/s_k is very close to 0.2. This is not surprising:

corresponding to each number that contributes to s_n, there are two numbers that contribute to s_{n+1}, each of roughly (but slightly less than) 1/10th the magnitude. (This explains why s_{k+1}/s_k is a bit smaller than 0.2.) Treating s_k from $k = 6$ onwards as a GP, we may estimate the sum $s_1 + s_2 + s_3 + \cdots$ (to infinity) to be

$$1.5 + 0.267316 + 0.052885 + 0.010565 + 0.002113 \\ + (0.000423 \times 1.25) \approx 1.83333.$$

(The term 1.25 comes from the relation $1+0.2+0.2^2+0.2^3+\cdots = 1.25$.) So we can be quite sure that the required sum is smaller than 1.83333, but not by much; the error is at the most in the 4th decimal place.

Chapter 14

14.4.1 Suppose that $\sqrt{3} = a/b$, where a, b are positive integers with no common factor. Then $a^2 = 3b^2$, so a^2, b^2 have the same parity (they are both odd or both even), and therefore a, b have the same parity. Since a, b are coprime, they are both odd. Thus, b^2 is of the form $4k + 1$. But this implies that $a^2 = 12k + 3$, that is, a^2 is of the form $4k' + 3$, which is not possible: no square is of the form $4k' + 3$. So this equation cannot hold good.

14.4.2 Suppose that $\sqrt{6} = a/b$, where a, b are coprime positive integers. Then $a^2 = 6b^2$, so a^2 is even, and therefore a is even, say $a = 2c$ where c is an integer. This yields $4c^2 = 6b^2$ or $2c^2 = 3b^2$. So $3b^2$ is even, therefore b^2 is even, so b itself is even. That is, a, b are both even, which contradicts the coprime nature of a, b. This contradiction shows that $\sqrt{6}$ is irrational.

14.4.3 Suppose that $\sqrt[3]{2} = a/b$ where a, b are coprime positive integers. Then, arguing in the familiar manner, $a^3 = 2b^3$, so a^3 is even, a is even, $a = 2c$ for some integer c, $2b^3 = 8c^3$, $b^3 = 4c^3$, b^3 is even, b is even; a contradiction; $\therefore \sqrt[3]{2}$ is irrational.

Chapter 15

15.5.1 From $r^2 = r + 1$ we get, by "completing the square", $4r^2 - 4r + 1 = 5$ or $(2r-1)^2 = 5$. Therefore, $2r - 1 = \pm\sqrt{5}$, that is, $r = (1 \pm \sqrt{5})/2$. Using the notation developed in the chapter, we see that the solutions are ϕ and $-1/\phi$.

15.5.2 We have

$$1+\frac{1}{u_n}=1+\frac{\text{Fib}_n}{\text{Fib}_{n+1}}=\frac{\text{Fib}_{n+1}+\text{Fib}_n}{\text{Fib}_{n+1}}=\frac{\text{Fib}_{n+2}}{\text{Fib}_{n+1}}=u_{n+1}.$$

15.5.3 (a) Consider any four consecutive Fibonacci numbers a, b, c, d. Here $c=a+b$ and $d=b+c=a+2b$. Consider the quantities c^2-bd and b^2-ac. We have, $c^2-bd=(a+b)^2-b(a+2b)=a^2+ab-b^2$ and $b^2-ac=b^2-a(a+b)=b^2-ab-a^2$. So $c^2-bd=-(b^2-ac)$. The implication of this is that, if u_n denotes the quantity $\text{Fib}_n^2-\text{Fib}_{n-1}\text{Fib}_{n+1}$, then $u_{n+1}=-u_n$. Applying the same idea repeatedly, we obtain $u_{n+1}=-u_n=u_{n-1}=-u_{n-2}=\cdots$, and ultimately $u_{n+1}=\pm u_2$. But $u_2=1^2-2=-1$, so we see that $u_n=\pm 1$ for all n.

(b) We use the result from (a). Let a, b, c, d be four consecutive Fibonacci numbers ($c=a+b$ and $d=b+c=a+2b$). Consider the quantity $ad-bc$. We have $ad-bc=a(a+2b)-b(a+b)=a^2+ab-b^2$. We already know from part (a) that $a^2+ab-b^2=\pm 1$, so it follows that $ad-bc=\pm 1$.

15.5.4 Let d denote the gcd of Fib_n and Fib_{n+1}. Then d also divides $\text{Fib}_{n+1}-\text{Fib}_n$, that is, d divides Fib_{n-1}. Since d divides both Fib_n and Fib_{n-1}, it also divides $\text{Fib}_n-\text{Fib}_{n-1}$, that is, d divides Fib_{n-2}. This can be continued all the way down to d dividing Fib_1. But $\text{Fib}_1=1$, so d must be 1. That is, the gcd of Fib_n and Fib_{n+1} is 1. So consecutive Fibonacci numbers are coprime.

15.5.5 The 1st two Fibonacci numbers are 1, 1. Considering only the remainders left on division by 2, the Fibonacci sequence takes on the following appearance: 1, 1, $1+1=0$, $1+0=1$, $0+1=1$, $1+1=0$, …; that is, 1, 1, 0, 1, 1, 0, 1, 1, 0, …. We see that the sequence 1, 1, 0 repeats. (It must, because each occurrence of 1, 1, 0 will be followed by another occurrence of the same cycle, because the next two terms are 1, 1.)

In the case of division by 3, the same idea may be used to give the following sequence: 1, 1, 2, 0, 2, 2, 1, 0, 1, 1, 2, 0, 2, …. On examining the sequence carefully, we find that the sequence 1, 1, 2, 0, 2, 2, 1, 0 repeats (the next two terms are 1, 1). Here the cycle length is 8.

15.5.6 The same idea followed in Problem 15.5.5 may be used here also, only we find that the cycle is rather long. The use of a computer shows that the cycle length is 60, the cycle itself being:

1, 1, 2, 3, 5, 8, 3, 1, 4, 5,
9, 4, 3, 7, 0, 7, 7, 4, 1, 5,
6, 1, 7, 8, 5, 3, 8, 1, 9, 0,
9, 9, 8, 7, 5, 2, 7, 9, 6, 5,
1, 6, 7, 3, 0, 3, 3, 6, 9, 5,
4, 9, 3, 2, 5, 7, 2, 9, 1, 0.

(Observe that the next two terms will be 1, 1, the same as the 1st two terms, and so the same sequence will be obtained once again.)

15.5.7 From the recursion $a_n = 3a_{n-1} - 2a_{n-2}$, we obtain the associated equation $r^2 = 3r - 2$, or $r^2 - 3r + 2 = 0$. This factorizes as $(r-1)(r-2) = 0$, so its solutions are $r = 1$ and $r = 2$. It means that the sequence of powers of 1 (that is, the sequence consisting only of 1s) and the sequence of powers of 2 *both* satisfy the given recursion. Therefore, the sequence whose nth term is $a \cdot 1 + b \cdot 2^n$ will also satisfy this recursion, for any constants a, b. So if we find a, b such that the 1st two terms of this sequence are 4 and 5, respectively, then we would have obtained the generating formula of the given sequence. So we must solve the equations $a + 2b = 4$, $a + 4b = 5$. These give $b = \frac{1}{2}$, $a = 3$; so the formula we have found is $a_n = 3 + \frac{1}{2}2^n$, that is, $a_n = 3 + 2^{n-1}$. (Check if the formula works!)

15.5.8 Proceeding as we did in Problem 15.5.7, we examine the equation $r^2 = 2r+1$, or $(r-1)^2 = 2$, whose solutions are $r = 1+\sqrt{2}$, $r = 1 - \sqrt{2}$. We now seek numbers a, b such that

$$a \cdot \left(1 + \sqrt{2}\right) + b \cdot \left(1 - \sqrt{2}\right) = 1, \quad a \cdot \left(1 + \sqrt{2}\right)^2 + b \cdot \left(1 - \sqrt{2}\right)^2 = 3.$$

The equations simplify to $(a+b)+(a-b)\sqrt{2} = 1$, $3(a+b)+2(a-b)\sqrt{2} = 3$; the solution is $a = b = \frac{1}{2}$. So we get

$$b_n = \frac{\left(1 + \sqrt{2}\right)^n + \left(1 - \sqrt{2}\right)^n}{2}.$$

15.5.9 (a) Similarity geometry reveals the answer instantly. The longer way is to compute the lengths of the diagonals. The longer diagonal has a length of $\sqrt{1^2 + \phi^2} = \sqrt{2 + \phi}$, and the shorter one has a length of $\sqrt{1^2 + (\phi - 1)^2} = \sqrt{3 - \phi}$. Now, observe that $\phi^2 \cdot (3 - \phi) = 2 + \phi$; for $\phi^2 = 1 + \phi$, and

$$(1 + \phi) \cdot (3 - \phi) = 2\phi + 3 - \phi^2 = \phi + 2.$$

Therefore, $\sqrt{2 + \phi} : \sqrt{3 - \phi} = \phi : 1$.

(b) This uses elementary trigonometry. We make repeated use of the following facts: (i) the diagonals of a regular pentagon trisect the angles at the vertices, so there are numerous angles equal to 36°, (ii) $\cos 72° = (\sqrt{5} - 1)/4 = 1/(2\phi)$. The details are left to the reader.

15.5.10 This will be left for you to tackle!

Appendix C

List of References

A reference list was given at the end of Part A (Section 6.1). The list given here is relevant for the chapters in Part B.

[1] G H Hardy & E M Wright. *Introduction to the Theory of Numbers* (ELBS).
This is a goldmine of results on elementary number theory and is relevant for Chapters 7, 8, 10, 11 and 15. It is also extremely well written and is highly recommended.

[2] L E Dickson. *History of the Theory of Numbers* (Chelsea).
This is written in a facts-only style but contains considerable material of value to those with an interest in number theory. (Relevant for Chapters 7, 8, 10, 11, 12 and 13.)

[3] John Conway. *"An old fact and some new ones"*, in Quantum, September–October 1990 (p. 24–26).
The article describes the 'curious procedure' presented in Chapter 9 (Sections 9.1, 9.2). It also suggests ways in which sequences other than the sequence of nth powers (for some n) may arise; e.g., by using the triangular numbers 1, 3, 6, 10, …, for the initial round of deletions we obtain the sequence of factorial numbers 1, 2, 6, 24, 120, …. (Relevant for Chapter 9.)

[4] Robert M. Young. *Excursions in Calculus—An Interplay Between the Continuous and the Discrete* (MAA).
This attractively written book has a vast amount of material on many topics in mathematics. It contains information on the Euler-Maclaurin summation formula, a powerful tool used to find approximate formulas for sums (e.g., for $1 + \sqrt{2} + \sqrt{3} + \cdots + \sqrt{n}$). (Relevant for Chapter 14.)

[5] *The Fibonacci Quarterly*.
This journal is exclusively devoted to studying the Fibonacci sequence and various cousins of this sequence. (Relevant for Chapter 15.)

[6] B Sury. *Induction—An Impresario of the Infinite* (*Resonance*, a monthly journal published by the Indian Academy of Sciences). This contains material on the technique of proof by mathematical induction. (Relevant for Appendix A)

[7] Shailesh Shirali. *Odd Behaviour of the Even Integer 2* (*Resonance*, October 2007). This contains many curious properties of the even integer 2, thereby justifying its description as "the oddest integer of all".

[8] Shailesh Shirali. *Infinite Descent – But not into Hell!* (*Resonance*, February 2003). This article showcases the method of descent, an important and powerful proof technique which may be regarded as a companion to the method of mathematical induction.

Index